Where the Hell's the Time Gone?

A LIFE IN FARMING

Best wishes, Tom.

I dedicate this book to my late wife Margaret for staying with me through the good and bad times, supporting me and keeping me on the straight and narrow.

To my daughter Amanda and son-in-law Phil, my son Mike and daughter-in-law Julie. To my grandchildren Dr Greg Thomas, Paul Thomas, Kelly Thomas and Amy Evans, and all the friends who helped me throughout my life in farming.

Tom Evans MBE
Retired Farmer, Champion Hedge Layer, Hedging Judge,
Shearer, Shearing Commentator and Judge,
Sheep Judge and After Dinner Speaker

Where the Hell's the Time Gone?

A LIFE IN FARMING

TOM EVANS

First impression: 2023

© Copyright Tom Evans and Y Lolfa Cyf., 2023

The contents of this book are subject to copyright, and may not be reproduced by any means, mechanical or electronic, without the prior, written consent of the publishers.

Cover design: Y Lolfa
Cover image: Ruth Rees Photography

ISBN: 978 1 80099 433 1

Published and printed in Wales
on paper from well-maintained forests by
Y Lolfa Cyf., Talybont, Ceredigion SY24 5HE
website www.ylolfa.com
e-mail ylolfa@ylolfa.com
tel 01970 832 304

Contents

1	Mother leaves home	7
2	School bullies	14
3	Breaking new ground	18
4	Chief cook and bottle-washer	23
5	The art of hedging	26
6	On the road	32
7	A place of my own	39
8	Farm fit for a wife	47
9	Running water, at last!	53
10	Struggling to make a living	58
11	Pastures new	64
12	Back to Wales	74
13	Lucky for some	85
14	The farming life once more	92
15	On the move again	99
16	Ty'n y Cwm	109
17	A growing business	116
18	Better times	122
19	A visit to France	127

20	Farming film stars	133
21	On the march in Brussels	138
22	Wedding bells	141
23	A victory for Welsh shearing	147
24	Another country	153
25	Stolen goods	158
26	Foot and mouth	161
27	Meeting the Queen	165
28	Time to slow down	169
29	More royal connections	172
30	Shearing Down Under	177
31	A health scare	186
32	Back to New Zealand	190
33	Home... and alone	198
34	Life goes on	203

CHAPTER 1

Mother leaves home

THIS IS MY story as it happened. I was born in Gladestry in 1943 and moved to Llwyn Tudor, Rhulen, which is a 146-acre farm my father had inherited from his grandmother. He had lost his mother at birth and was raised by his grandmother and was rather spoilt. When he was eighteen years old his gran bought him a new BSA 500cc Gold Star motorbike when most other boys were riding clapped-out old pedal bikes.

I was a very fast baby and was born seven weeks after my parents' wedding. At that time our house at Llwyn Tudor was cold and draughty. Later my two sisters Vera and Janet arrived, and for a few years all seemed to be well.

My first memory is of the big snow in 1947 and I well remember having to dig a way out of the house every morning because there was so much snow. When you opened the door, the doorway was full of snow, so you had to dig into the porch to get out. This went on for weeks. The worst snow came about 27 January. When it thawed I don't know, but the very bad weather had killed thousands of sheep that had gone down in a very weak state. The rivers were full of dead sheep for miles. We had about 300 ewes and reared only about twenty-four lambs around the hills of Llanwrtyd and Abergwesyn. The losses were bad. Several big flocks of 2,400 sheep were reduced to 400 after the storm. One farm that kept a herd of mountain ponies lost two-thirds of them.

All that was left were the dark bays, all the other coloured ponies had died. A disaster fund compensated farmers for their loss.

Of course, the war had just ended, so we were still rationing food, but as farmers we were never short because you could kill a pig, had hens for eggs, and the garden for vegetables. Farmers were under government control and had to grow a lot of corn and potatoes for them to keep the country fed. So it was the same then as it is now, people coming onto the farm and telling you what to do – and this was hard to put up with.

We did not have a tractor or a car until 1952. Ploughing was done by the War-Ag, which was government-run and was manned by servicemen coming back from the war and Land Girls brought from the towns. There were many stories of the capers they got up to. One story I heard was a farmer above Hundred House had the War-Ag to plough a field and then asked them to come and disc-harrow the field. A Land Girl came and did the job, but when he went to pay the bill he said to the man in the office that the bill was too big as she took longer than she should have because she spent most of the day 'peeing'!

We grew White's Victory oats on our land, which suited it very well. You could grow oats then without any spray and have a good crop. Our biggest problem was rabbits as our dry land suited them and they thrived at that time. Before thinking of planting oats, you spent a lot of time catching rabbits in gin traps, which were allowed then. Father caught hundreds. He and a man from the village who came to help would go out with an ash pole and collect the rabbits every morning and collect them in twos, put them over the pole and carry them, maybe twenty at a time. They could get up to eighty in one morning. These had to be paunched to get the 'guts' out so they would keep a few days whilst waiting for the lorry to pick them up. You never ever got on top of

Mother leaves home

the rabbits because, wherever you cleared, they moved in again. The corn harvest was important as we used a lot of oats to feed hens and sheep, also we sold some for seed.

The hay was cut with the horse mower. You would cut two or three acres, as much as you could handle by hand, turning, shaking it out and then raking it up with the horse and staking it in cobs ready to haul on the gambo. It was all hand work, but it was wonderful how it all got done in those days. It was mostly meadow hay, so it smelt lovely with all the herbs in it. When it was ready to haul our neighbours came to help – and when they were hauling, we helped them. It was the same with corn. We always had a barrel of cider in, ready for shearing and harvest. Sometimes we had two.

In 1951 something happened that was a blow to our family. Life was never to be the same again. My mother went to a whist drive in Cregina, taking my middle sister Vera with her. She never returned that night and when I woke up the next morning Father was in a flap and didn't have a clue what to do. In the middle of the night, he had walked down to Cregina to see where they were – they had disappeared off the face of the earth. It was several days later when Father found out she had gone off to Pembroke with the bachelor farmer next door.

My younger sister Janet was two-and-half years old, so Father had to do something. He took her down to my grandmother. She couldn't understand why her daughter would do such a thing. My sister was there for nearly twelve months. When the shock of what had happened finally sank in, my father nearly went mad. He was a wild man with an instant temper and, for me, living there on my own with him so depressed and mad about what happened was hard. It was worse at night because he would walk up and down the stairs, sometimes with a 12-bore loaded gun in his hand, shouting that he would shoot the man that ran off with my mother. I can assure you he would have done it if they showed their faces in the area. After a couple of weeks he calmed down and we had to get on with it.

Father was very friendly with the Eastoughs, a couple who lived at The Park on the other side of the Edw Valley. We walked over there one or two nights a week. They were very good to us and Mr Eastough gave Father some advice on what to do. He told him to get a solicitor to file for divorce and get custody of the kids, which he did. I will never forget that Christmas as the Eastoughs invited us over, and when we got there Mrs Eastough took me into the sitting room where there was a big settee which was covered with presents, mostly small things, but it cheered me up no end.

She was a lovely lady and Mr Eastough was an ex-army chap and smoked nonstop, which suited Father as he smoked sixty cigarettes a day himself, mostly the best Franklin, so the house was always full of smoke. Mrs Eastough kept pigs and always had plenty of meat about. She always had sows pigging, and on a cold night she would bring the piglets in and put them each side of the fire in boxes. Many times she would leave me cooking a big pan of sausages with the pigs squeaking, and two big Labrador dogs lying in the way, as well.

It was the beginning of November, so the salmon were up the Edw River, and I went fishing with Father. We liked salmon and we lived on it in November. When we finished, we went and caught some more. We sometimes poached trout as well.

By now it was 1957 and during that year I did not go to school much, but things changed after the divorce. A day I will never forget was at Swansea's divorce court. At that age I had to appear in front of the judge in his chamber. Fair play, he took his wig off so as not to frighten me, but I was still scared I can tell you. The judge asked me who I wanted to be with, my mother or my father. Of course, I had it drummed into me to say I wanted to be with my father, which I did. I

Mother leaves home

doubt if that happens today. Anyway, the outcome was that Father won the case of desertion and custody of us three kids and £1,000 in compensation costs.

It was good to get my sister back home, but this was just the start of seven or eight years of a rough life. The council found out and insisted Father got a housekeeper or we would have to go into a home. We had a few housekeepers, some were OK, but some were not so good. We had one who, when Father went to Builth on the bus on a Monday, as soon as he had gone, she would chase us out of the house with a broom! We were not allowed back in until just before Father came home at 4pm – she was crazy mad. When she went we had a local girl for a while, but she didn't stay, then a cousin from Gladestry came for a while, then went. So very often we had no housekeeper at all. By now I was older and could get my sisters ready for school and so we attended more often. Our school in Rhulen had seventeen kids in it, and our teacher Mrs Lloyd was a lovely lady who was there for all of my time in school from five to twelve years of age. She had three children, and her house was part of the school, so she spent more time changing nappies and feeding than she did teaching us.

Mother never forgot us at Christmas as she always sent cards and presents which Father threw on the fire in front of our eyes. We were not allowed to open them. I never saw my mother from 1951 to 1967 until I had a phone call that my gran was in Hereford hospital, so I went down on the Saturday. We were living in the Cotswolds at the time and when I went into the small ward to see Gran, there was a lady sat on the other side of the bed, and it took me a quarter of an hour to work out that she was my mother. She had changed so much in those years. On the way out of the hospital, my mother said that if I wanted to see Gran after she came out I would have to go to their farm, as Gran could no longer live on her own, so this was what happened. It was a hard day to go there and meet the man that had caused so much trouble, but I had to swallow my pride and go to see Gran, as she had been so good to me over the years.

At the age of twelve I was asked to join Edw Valley YFC, mainly because they wanted to hold hedging classes at our farm. Just after my thirteenth birthday I started learning the skill, and this was the start of a long hedging career. We were allowed time off school to go to these classes, which continued for three years – our teachers were champion hedgers. After our two years we were asked to hedge in a Hundred House hedging match. There were seven of us and I won that day, but I didn't win again for two years.

Out of the compensation money from the divorce, Father bought an Austin 12, a lovely old car which smelt of leather inside, and also an old Fordson tractor off the War-Ag, as it was being disbanded. This made life easier on the farm. All our implements had drawbars fitted instead of shafts, and we bought a Lister hay turner and a trailer plough. Father started to improve the farm; by now we had two cows which I learned to milk to supply the house. They were wicked old cows. Just when I had about enough milk, they would kick the bucket over and I would have to start again!

On the sheep side of our farm, we were producing Kerry cross ewes and ewe lambs to sell, and from the age of eleven or twelve I was always first up and around the ewes before going to school. I was pretty good at drawing out lambs that were stuck – you always had a few when you used a big Kerry ram on hardy Speckle ewes. I started this after Father, who had hands like shovels, took seven ewes to the vet's in Builth and ended up with six dead ewes and seven dead lambs because the vet also had hands like Father. He said to me, 'You have small hands, you should have a go.' So I did, and was soon able to get most things done.

The Kerry ram was widely used in those days and thousands of Kerry yearlings were sold in sales in Knighton, Kington,

Mother leaves home

Hay-on-Wye and Builth Wells, though most went down to Herefordshire to produce crossbred lambs. We sold ours as ewe lambs and a man from Herefordshire came every year to buy them. Morris Jones, The Rhos, our next-door neighbour, sold his ewes in Kington, and I spent some time over there holding ewes for him to trim. The auctioneer in Kington was very funny and if somebody didn't sell, he knocked them down to Dan Archer of *The Archers* on the radio!

CHAPTER 2

School bullies

AT THE AGE of eleven I took the eleven-plus exam and missed going to grammar school by one point – which was no wonder considering the amount of schooling I had! So, I went to the secondary modern in Llandrindod Wells by bus at 8am. Neither our housekeeper nor Father were any good in the mornings, so I got my food myself and off I went. Going to Llandrindod school was a shock to me, moving from a school of fourteen pupils to a school with 500, and I was also shy.

Very soon the other kids found this out and one boy started picking on me most days, so I had to harden up. I put up with this for a while, then noticed another boy some of them messed with, so I palled up with him. He told me not to worry as he would sort it – and he did. We were good friends through school and he used to spar with me in the gym and so I learned a bit of boxing. One day I found the trouble-making boy alone and said to him 'Hey Mister', and then gave him a hammering. He never bothered me again. In life I have found that if you stand up to bullies they don't like it very much.

On the farm things were improving. We were growing more corn and one job I enjoyed was stooking the corn after the binder. When I came home from school I would have tea and was stooking corn until dark. A chap from Rhulen used to help with jobs like this and was good company. Three or four weeks later he would also help to haul this corn in, as well as putting the sheaves in the barn. It was a lot of work

School bullies

putting sheaves side-by-side around the shed. Someone would be passing them to you and then you would put them tidy around the shed, starting around the outside and finishing in the middle, then starting around the outside again until the bay was full to the roof. The same method was used to load the trailer, and when you had loaded it you put two ropes over because the sheaves would slide off as simple as pie, of course. This job was made easier with the tractor, and from the age of twelve I drove the tractor in the field while they were loading it. This was hard because I had to use both feet to put the clutch down on the old Fordson.

I always looked forward to the thrashing drum coming in November to thrash the oats. Several neighbours came to help and to do the different jobs. You needed two on the bay to put the sheaves up on the drum and one to feed the drum. This was a dangerous job, with the drum spinning by your feet – one mistake and you would be killed. But you got used to it. When the corn had been thrashed, the straw came out loose at the back of the drum. This was a horrible job to handle. The straw had to be stacked in a rick ready to go back into the bay after thrashing was finished – it was a two-man job. The corn came out of the side of the drum into bags that had to be carried to the granary. The strongest men had this job as these bags would be fifty to seventy kilos in weight to lift off the ground – this would not be allowed today. As I got older, I did all these jobs, including feeding the drum, which I enjoyed. Father went to help the neighbours when they were thrashing.

Shearing was another job that neighbours came to help with. We had about 700 sheep in the early 1950s which included 200 wethers kept for the wool, and they lived on the hill all year round. The first job that had to be done was to wash the sheep. This meant walking them to Rhulen village to the wash-pool, which was a hard job as they didn't like being driven away from home. My job was to go in front

15

and make sure they went the right way, as there were a few turns. I would turn them and jump over the hedge and run and get in front of them again. It was hard getting the sheep to go down to the pen at the pool, as they knew what was happening. When it was finished and we got back on the road to home, they would go for home so fast I had a job to keep in front. Four or five days later they would be ready to shear. Washing got the grease out of the wool to make it easier for the hand-shearers to cut, but washing isn't required for today's shearers.

On shearing morning we would have the sheep in ready for when the men would arrive at 8.30am. When I was seven or eight years old, my job was treading the wool in the wool bags. A good shearer with a hand shear, like my father, would shear 100 sheep a day. The shearing went on until 7pm or 8pm with plenty of cider drunk during the day. Father would then shear for everybody that came to help us, when it was their turn. There seemed to be more time then than there is now. I started shearing by hand when I was thirteen years old, and with the new shearing machine at fourteen. I learnt to shear on the bench the same way my father sheared – which was around the rib – but later I went to work for Jack Hughes, the Cwm, and learnt a different style with longer blows down each side of the sheep which was much faster.

By the time I was fourteen I was fairly strong and could do most jobs. I was helping next door a bit as well to earn a pound a day. Morris Jones was good to work for and taught me a lot about how to get work done. He was one of the fastest men I worked with at any job and was also good company. He had started with nothing and did very well.

At that time we had a grant to improve the house and buildings under the old Hill Farm Scheme. We had an extension put on the end of the old house, one room wide, which meant an extra bedroom and bathroom, which was very modern then. We also had a fire grate that kept going

School bullies

all night, so it kept the house warm; a new roof too, and the walls were painted all round. It had been a very cold house before as there was no electric in our valley at that time.

Before we had a shed to store the oats, we used to store the best oats in the end bedroom, as they were worth a lot of money for seed, but this gave us a problem with rats. They would come through the stone walls and into the bedrooms at night! One night I was sleeping on the landing and, as Father and Mother came to bed, they shone the light on me and my pillow was red with blood. They got me out of bed and downstairs to wash me, and found two holes each side of my ear. A rat had bitten me when I was sleeping! I never slept with my ears out after that I can tell you. I put the sheet over my head and around my forehead with only my nose out. I very often woke up to find two or three rats on the bed. I used to kick my legs up and throw them off. I have never liked rats since.

About the same time we had running water installed. This came from a spring facing Rhos Farm, and had to be pumped 300 yards or so to get enough of a fall to run into the house. Inch-thick metal pipes were used, with a Lister pump down at the spring in a small shed that pumped once a week. I would take a gallon of petrol and grease the pump. Then we would get the pump started and, by the time it had used that gallon of petrol, there would be enough water for the next week. Before that we used to carry water from a well 300 yards down our lane. It was all uphill and carrying two buckets was heavy work by the time you got up to the house. There was only a rough lane up from the road, so we garaged the car down there as well.

CHAPTER 3

Breaking new ground

At fifteen I left school, having done alright. When I went to Llandrindod school there were three classes, the lowest being P and M the highest. I was put in P class. After six months we had exams and I went up to the O class which was better, and at the end of the year we had exams again and I went up to the M class. There was nowhere else to go, so I ended school in about the middle of the Ms.

Having joined the YFC at twelve years old, I was now getting involved with their YFC rallies. I learned more in the Young Farmers' Club than I ever did at school. Hedging training was something I never forgot and took part in matches all around Brecon and Radnor, sometimes winning, sometimes not, but I enjoyed every one.

Back home, Father was improving the farm by ploughing up ferny banks that had never been ploughed. We did one patch of fifteen acres facing Rhos Farm that was ploughed by a contractor, and we put a new fence along the bottom of it to separate it from the patch below. We also had a chap in with a crawler to disc down and get it level. We went out to sow it with grass seed, three of us sowing the bank by hand out of a small bucket each, and when we were about halfway through a thunderstorm came down over the hill across us – it was the worst one I had ever seen. Within minutes, hailstones like golf balls were landing and there was lightning across the ground. The clouds made it as dark as night, and the ground was moving under our feet. It was very frightening.

Breaking new ground

We had a job to get off the bank to get to shelter under the sheet with the crawler man. When the storm was gone, we went to look at the damage. The storm had washed so much soil, fern and roots down to our new fence it was covered to the top, and water was flooding between the pig netting and the barbed wire. In the middle of the bank there was a hollow where the water had run down; it made a hole under the fence which you could drive a car through! It had also washed deep ridges all the way along the bank and our seed had gone down the river with the lime and slag. We didn't try it again that year.

Twelve months later, we did the piece of land below. Now this was very steep. We had a bulldozer there to clear off the thorn trees and bushes, but it nearly ended in disaster! On the steepest part, the bulldozer man pushed a big thorn tree over and it got stuck on top of the roots. Unbeknown to him, there was a wet spot below the tree and when he pushed the trunk onto it, the whole thing, including the bulldozer, slid down the bank, gaining speed as it went, smashing through the fence into the wood below, and ending up in the brook at the bottom of the woods with the driver still sat in the seat. When we went down to him, he was as white as a ghost! He had careered through the trees all the way down the wood and never got hit by a branch or anything. He was a lucky boy, but I think he needed clean pants! He managed to get the crawler off the tree and drove down to the brook. We opened the weir for him to get through, but he got stuck in the sand and the bulldozer stayed there for twelve months before they got it out.

The next day they fetched another bulldozer with a fresh driver. There was a hedgerow down the middle of the bank which we wanted to get out and the driver got it all except for two huge ash stumps. We had a chap from Radnor come and blow them out with some gelignite, and we went with him to watch. The bottom one was the biggest. The first job was to split the stump to make room for the gelignite, so he bored a hole down in the middle, which was a bit rotten. Then he put a couple of pounds of gelignite in it, exploded it to split the stump about six to seven inches apart. Then he filled the crack with more gelignite from one side to the other. He said we had better move back a fair way and that we should put our arms across our heads in case any big bits reached us. He was grinning all over his face. He pressed the button. The ground shook, the stumps went up in the air about thirty or forty feet and then went flying down the bank. It was like a bomb going off, and pieces landed on our heads, but luckily nothing too big.

He blew up three smaller stumps and was gone, but not before saying if we wanted the bank harrowed after ploughing, he was the man to do it.

This work was being done under the old Hill Farming Grant scheme. Once cleared, Ernie Mills from Rhayader came to see it and he had a man to plough it with a County Crawler and a single farrow plough. On the morning he started, we went to the far end where about 150 yards of ground was not too steep. He started to work with a big plough that did not have a disc, only a coulter. The bank had never been ploughed before and had grown ferns six feet high. These were building up in front of the coulter and raising the plough off the ground. Huge heaps of old fern went through the plough, then it would try to go into the ground again only to fill up again. He tried a few furrows, but it did not get any easier.

I was watching and said to Ernie, 'What if I sat on the plough and gave it a firm kick when it started to fill?' This I did, and we had a full furrow all the way to the bottom. We did a few bouts and it stayed good, but by then I had a blister on my arse! I told Father to fetch me an old wool bag and some string to make a saddle on top of the plough. That done, we ploughed on for days and days. Once we had gone

Breaking new ground

150 yards, we had to go a quarter of a mile around to get to the top again. The patch was about twenty-five acres or more, and took a long while. When ploughed we had to get lime on there, as ferny land is short of lime. Ernie Mills' boss had a lorry and, when he rang Radnor quarry, he was told the boys were on strike. So Ernie said if we could bag it ourselves, we could have it. We went and bagged two good loads. There was a lot of work taking it out, as the bank was very steep to go across, so where we could we took it on our small trailer. We took one load and stacked it across the side of the trailer and, as we were going across the bank, the bottom wheel went in a hole. The load slipped across the trailer, taking it over and sending me flying through the air to land many yards down the bank and breaking the drawbar of the trailer!

A lot of this lime had to be pulled down from the top of the bank on a sheet of zinc with the front bent up, and the same with the 20-10-10 fertiliser. We got the lime out, but it took work and time. After that, I had to spread the remaining lime by hand. It was hot weather and, of course, I stripped off. Hot lime burns, so with all the sweating I burnt to my waist. I then sowed the 20-10-10. Afterwards, Ernie from Radnor came with his T20 diesel Fergie and hydraulic harrow to work it down. If you saw how steep it was you would think he was mad, but he did it for £35, which was cheap. We got it sown and we were glad to see it done. It was a good learning experience for a boy of fifteen.

That autumn we started draining some wet areas out of two good fields. This again was done by hand using clay pipes, as plastic pipes were not about then. It was hard work but very satisfying to see the water running out after you had finished. After we started growing more oats, we had a lot of straw. So Father bought some cattle every autumn and used it as litter and for eating. The cattle usually arrived with big horns that spoilt the look of them, so we got the vet to take them off. This was a rough job on strong cattle, and one year

21

we did it too soon and they got maggots in the hole where the horn came off. This had to be treated, which was a hell of a job. The vet gave us some blue powder to put in a bucket of water, and also a pump to spray it into their horns. As you can imagine, they didn't like it much. We put a halter on each one and I had to hold them while my father sprayed the holes. I was fourteen at the time and was fairly strong for a boy, but they would pull me about and one blue bullock would rear up and take me with him, with father shouting, 'Hold on to the bugger, boy.' We had to do it three days running, and every day they got wilder. There were twenty-seven of them and we walked them to Builth market to the great August sale. It was about seven miles and we had to drive them up through the town. They sold well.

That same year we sold the wethers off the hill as the wool trade had dropped. We had to plough a rough piece below the road and had a good crop of rape to fatten up our Kerry Cross lambs. Jack Yeomans from Abergavenny bought the lot on the yard for about £6 each.

That year also Father bought a new car, a Hillman Minx. This was a big improvement on the Austin 12. The registration number was FO9280. We had it for a few years. He also bought a Fordson Major diesel tractor, which was a good tractor and the one that I learnt to drive on.

CHAPTER 4

Chief cook and bottle-washer

BY NOW, HOME life was pretty hard. Father had become chairman of Hundred House Football Club, but this took him down the wrong road. On Friday nights he would have to make sure they had a team for the Saturday afternoon, which meant going to Builth to look for players. He would go to the Plough Inn on Bank Square, as there were a few boys there who played football. He would have to ply them with beer to get them to come, and he was spending too much time in the pub. Father smoked sixty cigarettes a day and the housekeeper, who used to go along too, smoked thirty or forty. The job was not good between them.

They were spoiling things. My sisters were growing up fast and needed looking after. Father and the housekeeper starting falling out a bit, so I spoke to my sisters about what was happening and they agreed, young as they were, that something had to give. We agreed to be as awkward as hell with the housekeeper for as long as it took, to see what would happen. A week or so later, I came back with my sister and there was one hell of a row going on in the house. The next thing we saw was the housekeeper going off down the lane with two suitcases! My sisters were over the moon because she had been very hard on them. I was big enough to take care of myself by then.

By helping to make that happen, it made more work for

me, as I now had my sisters to look after – to make sure they went to school, had ready for when they came home and, of course, washing by hand and ironing with an old flat iron. So any job I did, I had to be back early, although sometimes they just had to manage.

Losing our housekeeper did not stop our father smoking or drinking, but you learn to survive.

Leaving school at fifteen, I soon got jobs on farms in the area to earn a few pounds, and at £1 a day you took a long time to save much. But I enjoyed learning from different farmers and it was as good as a college for me.

In the autumn, the salmon came up the Edw and I went fishing on my own and soon was able to catch enough to keep us going, as we all loved fresh salmon. A lot of fish coming up the river were seven to fourteen pounds, and so a nice, cooked salmon would make three feeds for the four of us. I learnt how to dress, clean and cook a fish. We cut the salmon into rings, then dipped them into flour and salt and pepper, then fried them in a hot pan. They didn't take long to cook. I always cooked the biggest piece first and then placed them around the plate, putting them on top of one another until the tail section was on the top. There is nothing better than five or six rings of salmon and some bread and butter for a good feed. The keepers were pretty hot, but I never ever got caught!

In 1960 the biggest fish I caught in the Edw was twenty-three pounds, up on Graig-yr-Onnen bottom, he was a beauty! Morris Jones, The Rhos, had asked me to get him a good fish for someone down in Lyonshall who bought cattle off him, so he was very happy.

In Hereford a couple of years ago I was talking to some farmers from that area and one asked whether I was the chap who caught that salmon. I said I was. It would be over fifty years ago and they still remembered. I did catch one in the Wye about the same time, he was thirty pounds. It was a cock fish caught on Dolagored Farm near Builth Road. We were hedging down to the river, and I saw him in the daytime and went back that night and caught him. He was a monster.

CHAPTER 5

The art of hedging

THAT SPRING I was sixteen and working around the neighbouring farms. I helped Jack James, Penbank, on the opposite side of the valley to us, to lay an old hedge which had not been done for fifty years. Jack had made a heap of ash stakes below the road, and I was asked to go and sharpen them. It was during the last days of February, a beautiful morning, and the sun was out. I went over with my axe, which was pretty sharp, and before long the sweat was dropping off my nose, so I took my shirt off. By 11am it was really hot, so I took my vest off as well, I was naked to the waist. By 5pm I had sharpened that heap of stakes – I was a bit stiff the next day but it was done. A few days later we were hedging right in the east wind and it was too cold to stand outside, but it was OK hedging because it was a warm job. I hedged with Jack for about a fortnight, and he taught me a lot about how to manage a big hedge.

In early May we were going to plant 11,000 cow cabbage plants on a good piece of land facing Cregrina village. We had done the same thing for a couple of years and used them to fatten cattle and lambs, so we had to put baler string down to get a line up the field. I had made a pouch to carry the plants, so it was easy to take one to plant with the spade. You pushed the spade into the ground and pushed the spade away and shoved the cabbage root down behind the spade then pulled the spade out and put your foot on the bottom of the cabbage to firm the soil. Once you got used to it, you planted them pretty fast. After two hours I had pain in my side which I thought was a stitch but carried on until dinner. I couldn't eat any dinner, but I had a couple of aspirin, a cup of tea and went back to planting. It took me three days to plant them all.

At that time I had broken the back of the garden work at home, taking out five feet of nettles. I grew all sorts of plants ready to plant out, so I always took a heap of plants out to the field to fill any gaps where plants had failed. That autumn was a good one and it felt satisfying to go to the fields on a Sunday morning and come home with a cauliflower or a Savoy cabbage, and later some sprouts, which grew well in the field. Nobody grows cabbage these days, but they made a lot of feed, as one cow cabbage would fill a wheelbarrow.

That year I did a lot of shearing around farms, and at the end of June and early July, Garney Morris rang up wanting Father to go shearing. We had hay to see to, so he said he couldn't go. Garney asked could I shear and Father said yes. So he asked for me to go – he was desperate as everybody was on the hay.

So, Jeff Langford from Hundred House and I went. We were both sixteen, so it was a big day for us to join the shearing gang. We went down to Coed Owen Farm, Cwmtaf, above Merthyr Tydfil. I was put to shear between two old hands at the job. It was a hell of a hard day with Billy Lloyd one side doing two to my one sheep, and Milwyn Harley the other side, doing one-and-a-half to my one sheep. I enjoyed every minute, but I was knackered that night. That was the start of my long shearing career and I will say more about that later on.

A couple of days after going shearing, I was turning hay all day and in the afternoon sun the pain I had planting cabbage plants came back. By 6pm I was bad. I stuck to it for a while, but then I worked out what the pain might be, so I went to the doctor. He was a Scotsman and he asked me, 'What's wrong, laddie?' and I replied that I thought I'd got appendicitis. He asked me to lie on the couch, put two fingers on my stomach and pressed down and I nearly screamed. He said, 'You are quite right, young man,' and told me to go up to the hospital in Builth and book in; he'd be there soon. I went and the nurse took me to my bed, pulled the curtain round and I got undressed. She said she'd have to shave me down below. This was a new thing for me to have a woman pulling my old man about to shave around him, but she was very good! By 7.30pm I was in the operating theatre having the operation. Next morning, the doctor came to see me and told me that one more bump on that tractor and I could have been in hospital for six months. I was a shy boy going in, but I wasn't shy coming out! Those nurses did a great job.

Farming was changing a lot then – we stopped growing corn and buying cattle and increased the sheep and changed the breed. We started using Radnor rams to breed ewes that would live out on the hill where the wethers had been. We had hill rights to use, so instead of buying ewes in, we kept ewe lambs by the Radnor rams and sent them on tack down to Worcestershire for the winter.

I continued to work around the farms, still on £1 per day in the autumn, pulling swedes and turnips for different farms. It's a cold job on the hands and you get stung by the nettles, so when you washed your hands at night your hands would sting like hell – you didn't feel it when they were cold. Then I started hedging, piecework, this was much better than day work. My first contract was for Elwyn Evans next door. He didn't like hedging, so no hedging had been done there for years. They were big hazel hedges – all axe work but good hedges. I took them on £1 a perk, which is seven yards, so for every seven yards I hedged I made £1. So much better than £1 a day! So, I got stuck in. I had a good Elwel axe that, when sharp, would shave your arm. When doing big hedges like that you had to have a sharp axe. I could soon do thirty-five

The art of hedging

yards a day which is five perks, so instead of £1 a day I was on £5 a day. I hedged all winter when the weather was dry. Doing piecework hedging helped when I hedged in a match because you learnt how to get on with the job. I started to win a few matches, but the classes were big in hedging matches then, so to win you had to be good.

In the following autumn I hedged in a Rhayader match. I drew a piece of blackthorn, which I like to hedge with, and did a good job except I was too near the top of the ditch with my hedge, and to cut it off would have spoilt the look of it. The judge that day was Percy Goodwin. He came by and said in his Radnor twang, 'Thee can't have it today, boy, you be over the ditch', so I ended up third and fourth divided with John Lloyd, Llanshiver. Anyway, John Davies won, and Donald Mills was second. My mate Jeff Davies was hedging in a different class. They were judging that with a Tilly lamp, and Jeff was not happy he had second.

The following Saturday we were in Erwood on the side of the main road, past the White House below Builth. There was a good turnout of boys again, and Father took me because I still couldn't get a driving licence. When they made the draw, I drew a piece by the gate on the side of the road. It was a hopeless piece of nothing but some little blackthorns and a bunch of briars. There was one ash tree four inches through, with the top cut out by the telephone people, and in the middle was an iron hurdle. I spoke to the stewards about it and Father told me to make up my bloody mind if I was going to hedge or not – 'If you aren't, get in the car and let's get home.' He was pretty mad, but Tom Abberley, one of the stewards, told him to hold on a minute, they would see what they could do. They came back and said if I would hedge it, they would bring all the good hazels I wanted out of the wood above, so I said I would do it.

I got my stuff out of the car and Father went home. I got stuck in and put a nice ditch up and started hedging. The

29

stewards fetched me wood from above and it came down alright using the briars for backing on the roadside and the blackthorn on the field side. I managed to get the ash pleached. I got it cut down and started topping off, and by 3pm it was looking OK. Percy Goodwin came down through the gate – he was judging again – and he had a look down my hedge and said, 'Thee has got better job today, boy', and I won against a good class including Donald Mills, who was very hard to beat. I was pleased to win from a bad piece.

I have got about ten first prize cards for every match in the area. There were seven matches then – Rhayader, Llanbister, Builth, Erwood, Hundred House, Radnor Valley, Talgarth & Rhosgoch.

This year again, I did more shearing, getting some contracts of my own and improving as I got older and stronger. The machines were getting better, wider combs were available and combs with one tine turned out; it all helped to get the wool off faster. I had a motorbike by then and would ride to Llanyre to shear for Garney Morris. I remember going one morning about 20 June and the ground was white with frost. The fern on Graig-yr-Onnen common was black when I came home that night – it was very cold on the bike that morning.

Shearing with Garney paid £5 per day, and he supplied the machines. It was good pay when the farm wage was £7 per week, but you earned it. By then balers were becoming available, so getting the hay in was much easier. Most baling was done by contractors and we had a chap by the name of John Cornish from Erwood to do ours. He also had a stock lorry and kept pigs.

Once our hay was in, I helped Morris Jones, The Rhos, to haul his hay as he made more than us because he had a lot of cows to winter. He had a workman called Austin Davies who was the strongest boy about. A bale was nothing to him, so he was good to work with. Later, another neighbour across the valley asked if I would go and help him cut his corn with a scythe because his field was too steep to cut with a binder. I had never used a scythe but soon learnt how to use it and we cut the field between us while his sister made the sheaves. We stacked what we cut every night.

Some years we grew swedes and turnips. They had to be hoed, which was a steady job, but I enjoyed hoeing and also helped Morris Jones hoe his swedes. In the autumn I helped to pull the swedes as well. Many times, hoeing at The Rhos, there would be Morris Jones, his wife, Austin the workman, Adrian their son and me. Now I am the only one left alive – Morris and his wife died with cancer, so did Austin the workman, and Adrian collapsed on the yard and died suddenly. So very sad.

CHAPTER 6

On the road

By now I was old enough to get a licence to drive a car and a lady down the road, Blanche Lewis, gave me a few driving lessons. She had been driving about two years and after two or three months I had my test and just managed to pass. Blanche was very pleased. The Lewis family were very good to me through the hard times. Bert Lewis had a straight leg and walked with a limp, but could do most things. They had a smallholding of twelve acres. They kept two cows and calves – the cows were milked on one side and calves sucked the other. They made their own butter and fed the buttermilk to the pig, which they would kill for meat.

Father used to go down to help with killing the pig, and when it was cut up he always came home with some part to eat. The pig meat was salted on a salting stone in the pantry. The salt came in twenty-pound blocks which had to be crushed with a rolling pin so that it could be used to cover the joints of the meat. There would be six large joints each side of the pig – the ham, the thin flitch and the thick flitch, and then the shoulder. There would be about seventy kilos on each side. A layer of salt was placed under the joints on the stone, then salt was put all over the joints with saltpetre in the ends of the legs to stop flies getting into the meat. If you had any small pieces of meat this would be eaten pretty soon; that was placed on top. Much of this meat was very fatty, especially the flitches between the shoulder and the ham where there would only be one layer of lean meat. This was what we ate in the week, with shoulder or ham only at the weekend. There were five in the Lewis family, so they would eat one pig a year. Bert had a very good garden and in wintertime he would shoot rabbits and hares. They liked salmon as well. Mrs Lewis was always very good to me.

Having passed my driving test, I used the money I had saved to buy an A30 Austin van for £120. This was very good as I now had wheels, so I was independent and could go where I wanted. I could also travel further away for work, hedging, shearing and, of course, courting my future wife Margaret over in Rhosgoch.

We used to go out every weekend somewhere. At the end of a dance we would make our way home and find a pull-in somewhere to do some canoodling. The van was cold, but we kept warm enough. I knew every pull-in for miles around Rhosgoch and Clyro!

I was still taking part in hedging matches and travelling further afield. I was hedging down in Esklyside and Leominster and gaining more skills, but I never had much luck in matches there as the hedges were long and plain, not suiting my style. Me and Jeff Davies went to Leominster one year and didn't win. One our way home we went to a dance in Clyro. Jeff said we could call by his uncle, a Mr Layton, and change our clothes. I found I had forgotten my shoes, but Mr Layton said I was about the same size as him and lent me his best shoes to go dancing. Margaret and I took them back on the Sunday.

That autumn I caught my record number of salmon for a single season. It was a good year and I caught 106, but my catch was nothing compared to two chaps down the valley who had caught 460 and were selling them in Merthyr. One night, a chap phoned me up and asked if I would go fishing with him. I told him it wasn't much good as the river was rising fast from the rain the two days before, but he came anyway. Down we went and I told him the only place we might see some was the Sally pool ford, so that's where we went. I shone a torch on the river and said to him, 'Look there!' I counted twenty-six fish right in front of our eyes. I told him to get a bag open, and we started catching them. After a few minutes he shouted to stop as we couldn't carry any more. He had filled two bags with lovely fresh fish. That was the best night of the season but we had a lot of fun salmoning that winter and I also did more hedging to earn money to keep my van on the road.

We were pretty active in the Young Farmers and decided to put a team in the Radnor Public Speaking competition. We practised hard. The team consisted of me, Jeff Davies, John Davies Llangoverous, and Cath Morgan, a student who was working at the Forest Farm and who was very good at speaking – she lives in New Zealand now.

The competition was in Rhosgoch, so we arrived by 7.30pm. We booked in and drew our place so we knew when we would be on. We also drew our subjects; we drew 'Getting up in the morning and growing potatoes on a Welsh hillside' which was interesting to say the least. We were drawn about two from last so had a long time to get nervous. I had taken a pound of sweets and we ate them all by 8.30pm. Jeff said, 'Let's go to the pub,' so we went down to Painscastle and had three or four Nut Browns and went back. This provided us with a bit of Dutch courage for the competition. Jeff Davies was the speaker, I was the proposer, John seconded, while Cath Morgan summed up and thanked everybody. We were on good form and managed to make people laugh with some comments and this helped us to win on the night. Taking part in that competition helped me a lot later in life.

My sisters were growing up and Vera, my oldest sister, had left school, which was a great help and made life easier for me. She was also good around the farm. Both my sisters had joined the Young Farmers, which also did them good. In early June we decided to put a YFC team in the rally. I was given a few competitions and Jeff Davies and I were to do the hurdle-making, which we won. I judged the Welsh Black cattle and came third, and I identified grasses and was about sixth. I was also asked to do the rolling of the rick sheet, so I practised in the barn at home with a clock because it was about timing as well as tidiness. Anyway, I came third.

That day we ended up winning the rally and had a few beers that night! We did have a very strong club at the time with a lot of talented girls competing in their competitions. Our club leaders were very good and encouraged the youngsters to get involved. The Young Farmers is a wonderful movement for young people to learn how to mix with others, and it is a very good place to find a future wife. Many have found wives that way, including me!

That summer I did more shearing than before and was getting pretty good at it. When I went with Garney Morris to shear, I was put next to Matthew Price who was the best bench shearer I had worked with. He was very fast and very clean and would keep going from morning until night, so I learnt a bit off him and improved to the point where I could keep up with him most days. He enjoyed me pushing him along too; he wouldn't let you go past very often. We travelled far and wide that summer, up to the hills of Abergwesyn, to Pentwyn, Nantstalwyn and Dolgoch. They had a lot of sheep then, but there are not as many now since a lot of the land has been planted with trees.

The sheep at Nantstalwyn were shorn by hand before machines were brought in, and my next-door neighbour told me there would be 106 hand-shearers there to shear the flock. Hand shears were first made in 1730 by Burgen and Ball from Bristol, who still make shears today for export to South Africa. Most of the sheep – over 10 million – are still shorn by hand out there as there is no power in the hills. On many farms in the Welsh hills, we would be shearing by machine in one shed and, around the back, the ewe lambs would be shorn by hand.

Another big flock was Nant y Maen on the road to Tregaron. When you arrived in the morning the fields below the house would be white with sheep, you could not see any green. We used to be about twelve shearers and Garney always sent the best men there because of the number of sheep. By about 5.30pm to 6pm they would be gone. Large areas of that hill have been planted now, but there is still a good flock of ewes there today.

We also sheared for Roger Davies, Hafdre, who at that time had 2,500 sheep. Garney also had the contract to shear at Pwllpeiran, Cwmystwyth. Back then this was an experimental farm. When you sheared there, they had students weighing every fleece, so at the end of the day you had a total for each shearer. Only Garney knew the number as he paid a flat rate of £5 per day. Matthew and I went there one day, and he told me that people were saying that the boys using the new Bowen-style from New Zealand were shearing more sheep than us. Ken Powell was shearing with us too and they said he was much faster than us, so Garney asked us to try harder to beat him. I said we would. Ken took off flat-out in the morning and I stuck to him until dinner and dropped just one ewe. After dinner, they fetched a bunch of Welsh ewe lambs to shear. I hated shearing lambs, and lost seven to Matthew through the lambs, but when we got back on the ewes I held him to the end, not losing one ewe. When the numbers were totted up, Matthew had shorn 220 and I had done 212. The nearest to us was Ken Powell, but way back on 186, so we proved the talk was wrong!

I had some hard days with Eddie Morris. We sheared his uncle's sheep in Abbeycwmhir one year – they were big Kerry ewes, and by the end of the day I could hardly stand. We finished them and did 186 ewes each that day. Eddie's knee used to come out.

Sometimes I would finish his ewe and he would massage his knee, clonk it back in and then get back to shearing.

On the road

I sheared with several different men, but I enjoyed shearing with Garney Morris most. He always treated us well and the men shearing with him were great characters. You always had some fun when you were with chaps like Jack Lloyd, Tom Richards, Brian Jones, Denis Hardwick, Viv the Vaynor, Billy the Swydd, Billy Bevan, Michael Gayther, George Hardwick and the Davies boys from Merthyr Cynog, Goronwy and Gwynfor, who both emigrated to New Zealand to shear. I had good days shearing with these fellows, never a dull moment.

Autumn brought more swede-pulling and soon it was hedging time again, so I could earn enough to keep my van going. Petrol was a bit cheaper than today, five shillings and threepence, 5s. 3d. (26p) per gallon for four-star and four shillings and eleven pence, 4s. 11d. (24p), per gallon for three-star. Farming was changing – most farmers by now had a tractor and a car or a van, but roads were still very quiet in 1960, not like today.

People were trying different breeds of sheep, the Beulah Speckled Face Sheep Society was formed in 1957 and soon had sales in Builth Wells. The breed was taking over from the Kerry and the Border Leicester that were used on Welsh ewes to produce the Welsh Half-bred. The Clun Forest was also used on the Welsh ewe to produce ewes for fat lamb production down in Herefordshire. Father brought a Beulah Speckle to put to some of our Radnors which didn't work very well, so we put them to some hardy Speckles which worked better.

During the winter, when it was too wet to hedge, I went to do building work for Jack Lloyd – roof-maintenance, putting bathrooms in and the sewage pits to go with it. Farmers were starting to modernise their houses and were fed up with going down the garden to the loo. The best outside loo I ever saw was on a farm in the Elan Valley by the side of one of the dams. It was built over a small stream, so it was an

37

automatic flush and the stream went into the reservoir and it helped to sweeten the water in Birmingham!

I also started getting asked to do some fencing on farms because there were now so few staff on the farms. It paid OK, about £3 per roll. I could put up two rolls a day, no problem, so it was another job for summer and autumn.

I did a long length of hill fence at Hondon, Rhosgoch, which was handy. Bert Lewis came along to help on a couple of days. He knew a bit about how to do the job tidy, so I learnt a bit from him. In the autumn of 1962, I was working for Ted Breeze, The Pant, Rhulen, as he was crocked up with a bad back. One day he said he was going to sell up. I asked him what he wanted for the place, it was fifty-four acres. He said he wanted £5,300, so I offered him £5,150 and he said if I could find the money, I could have it as he'd had enough.

CHAPTER 7

A place of my own

AFTER THE CONVERSATION with Ted I went home. It was about 9.30am and Father was still in bed, so I went up and told him what I had done. He said I was 'off my bloody head', and that he couldn't help me with any money. I was a bit shocked as I had not had anything to do with the business, so I didn't know how the finances were. Anyway, he said 'You had better go and see your grandmother in Gladestry.' So I changed and went over there and told Gran what I had done. She said I had stuck my neck out, but if I waited, she would get changed and I could take her down to the bank in Kington. I dropped her outside, and after a while she came out and gave me an envelope, saying that was all she could spare. In the envelope was £500 in notes. I thanked her very much and went home – I now had enough for a deposit. Father then told me to go and see a distant relation who had inherited two cottages and some land from a couple of aunties. I'd only met this man a couple of times, but anyway, over I went that night to see him and told him what I was after. He said my father was right and he had got some money put by – he would lend me £3,600 if that would help. The rate of interest was eight per cent at that time. I thanked him very much.

Next morning, I went to Trevor Evans, the solicitor, in Builth and he said he would find the rest of the money, no problem. The next day, he was on the phone to say he had got the other £1,200 on a second mortgage at eight per cent again, so I was on my way. We arranged a settlement day and got it all together. The morning I bought the farm I had £25 shearing money in my pocket to stock the place. I'd not taken any money from home since I was fifteen, so Father bought me eighty-six hardy Speckle ewes from a Llandovery sale and gave me twenty-seven Radnor yearlings from home.

I went to open a bank account and asked the bank manager for a small loan. He was not happy but did allow me a loan of £500 working capital. I went to Rhayader and bought sixty little Welsh ewes from Allt Goch, Llanwrthwl, for £2 apiece. They were a good buy; they lasted a while and were good mothers, so I was stocked up. I also bought a Suffolk ram off Garney Richards for £20 and a Hardy Speckle from Rhosco Jones for £12.

The barn at The Pant was full of hay. I mentioned it to Ted Breeze, and he informed me that West Brecs (West Breckonshire Farmers) in Builth had been to see it and they offered him £17 a ton. I said I couldn't offer that as it was too dear, so we fell out. It was way over the top anyway. I went to Builth market on the Monday and saw Bob Williams, a salesman for Passey Nott, and asked him what price he could get on a good load of green hay for sheep. He said I could have a load off Whites of Worcester for £13 per ton. So, I ordered a load and then enquired about white maize. He said that was £26 per ton. I ordered two and told him I wanted it by mid-November. He also informed me that, as I was starting out in farming, he didn't want the cheque for the bill until September to October next autumn, which was a big help. Most firms were very good for credit in those days.

I also brought a Ferguson tractor, a T20 TVO for £60, one of the first tractors Mike Hughes, Smithfield Tractors, sold. She was like new, with new tyres front and back. She was a lovely little tractor and pulled well for her size. I also bought a transport box for taking the hay out in the winter.

In 1962 I was back on the hedging and fencing in the autumn to help with the finances. When Christmas arrived,

A place of my own

things came to a halt on Boxing Day. It started snowing and it snowed and blew for three days. It blocked everything up, all the roads in the area were buried in snowdrifts ten to fifteen feet deep. You had to walk everywhere.

I had got the sheep into the yard at The Pant and opened every door into the cowsheds and the stable so they could get out of the cold. And cold it was. When the snow stopped, it froze every night down to twenty degrees below, and the wind in the east made it worse. I started feeding the ewes in the orchard and after a few days it was just a single sheet of ice, but that was good for feeding my white maize on and they didn't waste a grain of it. My 170 ewes had a big bucketful morning and night; good hay in the morning, poor hay at night. I had left some poor old bales which came in handy.

After a few days the council foreman rang up and asked what I was doing. When I told him that I walking to The Pant to feed, he asked did I want a job. I said yes and he asked if I had a good shovel – I said I did! The council was desperate for help to dig the road open from Aberedw to Rhulen. I asked when he wanted me to start and he said the following morning. I was to walk down to meet the lorry that was starting up from Aberedw – so that's what I did. It's five miles to Aberedw and I walked nearly all the way before I met them. I cut snow every day for nine weeks, including Sundays, for £7.50 per week and was glad to have it. We opened the road to Rhulen in a few days and after that we had to keep it open. The east wind blew the snow back in and there were plenty of mornings when I would walk halfway to Aberedw and then have to dig my way back up to Rhulen. My day was made up of a walk around The Pant to feed the sheep, then meet the lorry and later walk back by The Pant, feed the sheep again and walk home. I started out in the dark and by the time I got home it was fully dark again.

After the snow had been down for about a month, the farmer above Rhulen was running out of fodder for his sheep, so we had to open a way to get feed up there. This meant opening the fence into my field to save digging the road out, as this was blocked by snow ten feet deep. Going across my fields where the snow had blown off would be easier. When we got to Rhulen village, the road was open for a bit until we got to the Noyadd Pitch. This was about 300 yards long and was blocked with snow fifteen feet deep. With high ditches both sides and a hedge on the top, there was no other way to go. Luckily, the farmer had his tractor and box parked at the bottom, as he was afraid of being blocked in, so we borrowed his tractor and hauled the snow out into the field below the road. It took a few days. Once we cleared that, we had to dig here and there to get across to the house.

The next job was to get the hay near the telephone kiosk at Upper Rhos out to the Pentre. The farmer had five tons of hay brought that far, so after walking home for tea I returned and we started taking the hay up in a transport box, a load at a time. About fourteen bales would fit on and we had to put a rope around it to keep them on and, of course, it was dark by then. I had to walk in front with a Tilley lamp, as the others had no lights on their Fergie. We would do about three loads a night. After three nights the farmer said that would do for a while.

About ten days later, the farmer was out of fodder again but by then it had snowed more and blown into the Pitch, so we had to dig it all out again to get more hay in. I think we'd dug it out four times by March – we had a mountain of snow in the field, you could drive up on top and turn around to tip the snow over the end of the heap.

After about three weeks, the snowdrifts were frozen solid. So I said to the boys digging the snow that I was wondering if my A30 van would come down over the drifts to the kiosk. They said they would come with me and help if I got stuck. When we had a bit of time, we walked up and managed to get the van started and set off down over the drifts. It was a bit hairy in places but I kept going all of the way down to Rhulen school. From there up to the kiosk was a steep pitch, so they helped me to get up there. I now had wheels which was great.

I was finally able to go to see my girlfriend who was stuck in Llandrindod Wells where she was staying in digs and working at the council offices. She was very pleased to see me.

One night I went out and it was blowing more light snow. On one short length of road, about a hundred yards, the snow was blowing through the hedge and I thought it might give me trouble getting home later. When I returned, all was OK as I came back up the road from Aberedw, until I turned into the straight where I could see the snow still blowing through the hedge. I thought I would give it some welly and hope to get through, but I had a shock when the van went straight in up to the windscreen! I had a job to get out and had to walk quarter of a mile in shoes to get my wellies out of the kiosk. I had my legged pulled the next morning by the rest of the boys cutting snow!

We still had no electricity supply in Rhulen, so on a Saturday night I would walk over to The Rhos to visit Morris Jones and family. They had electric put in in 1962, so they had a telly and I liked to watch *That Was the Week That Was* with David Frost, which was very funny. We very often watched telly until 11pm. By then, I would be ready for home and Morris would ask me to hang on a minute and look around the cows in case anything was calving. Sometimes there would be, so I would give him a hand to pull a calf before heading off. At that time of night, just before midnight, you didn't need a torch to see. It would be freezing so hard it was like daylight in the moonlight.

I used to carry a gun home from The Pant after feeding and sometimes shoot a hare which we liked to eat. One night I was stood in a good place and managed to shoot two coming down onto some clover. I turned around and took the hares down to Upper Rhos to the Lewis family. They were very good to me when times had been tough. They were very pleased to have hares and I was told to call for tea the following night. I did and Mrs Lewis had cooked them lovely, so I enjoyed a good tea.

It kept freezing and being snowy until 12 March 1963. The wind had been in the east for weeks but that morning turned to a southerly and the temperature went up and the thaw began. What a relief it was! By then my fodder was getting less – I had used every bit of old hay from the tolent (loft) above the cowshed. Some of it had been there for years, along with some mouldy clover hay out of the Dutch barn. I had some sheep cake delivered to the kiosk, so I carried a bag a day to feed the sheep as lambing was getting nearer. The roads would not be opened for a while. In fact, the road from the kiosk by Upper Rhos to Cregrina did not open until mid-March and by The Pant until even later, but we managed.

I started lambing about 25 March. By then the weather was sunny by day but still frosty at night, so when the Suffolk cross lambs started to come I put them in the Dutch barn for the first night and they were fine after that.

I had a good lambing considering the winter. I lost seven ewes but still reared about 175 lambs; the Suffolk crosses were the first to be reared in the area. The next-door farmer came by one day when I was putting them down below the road. He was an old-fashioned old boy and he told me 'One flake of snow and they will be dead.' A few years later he had Suffolks himself on his farm.

I was not able to do any hedging until the first week of May because of the snowdrifts, but did some until mid-May, and fencing afterwards.

In May the electricity board brought a connection up to Rhulen, so I had to get the house and building wired, which cost about £100 plus £50 connection fee. It was a big

A place of my own

improvement. There was no water at The Pant so I was on the lookout for a spring. I found a very good one out on the hill above a little dingle called Pig Tail, it was right at the top on the left-hand side. I asked my solicitor how to get permission and he said that I should get in touch with the lord of the manor on the common, who were the Legge-Bourkes, Glanusk. I contacted them and was given permission to take water from that spring, but first I had to make some money to pay for it.

I put a few fields up for hay and had a good harvest which would be a good help the following winter. I did my usual stint shearing for Garney Morris and I was pleased to see some good young shearers coming out. Most of these young boys were shearing Bowen-style (which involves using the non-shearing hand to stretch out the skin on the sheep to produce an evenly shorn fleece). By this time Garney was shearing over 100,000 sheep with over thirty shearers out on some days from 25 June onwards.

We were shearing a lot of sheep around Llangammarch Wells for Bill Price, Aberceirios; Eddie, Llwynbrain; and Cefnllan. Garth House had a flock on Epynt at that time and Llwynowen, Beulah, had a hill flock which they do not have now. These were good sheep to shear.

One year, at Aberceiros, when they got the sheep in to shear, they found 250 to 300 were missing. Later that morning they found the sheep on a ferny bank after it had been raining, so they brought them in. They asked Garney about coming back the next day, but he said he was booked every day for three weeks, so could they shear them wet? Bill Price was worried about the wool, but anyway he rang the Wool Board in Builth and they said they would take the wool if Bill hung the wool over some ladders to drip for a day or two. In came the sheep and Garney went to the pub and brought twelve flagons of beer out, telling us to get wet inside before we got wet outside! This was the only time I have shorn wet sheep. The first one I did made me wet all down the front. They were cold against you but they were lovely because you didn't get any grease on the back of the comb. We didn't hold them long, I can tell you. It was coming off fast and furious! Garney came by and told me, 'Take your time, boy, you'll have a leg off in a minute.' Then he carried on past, laughing as he went. Jeff Davies was next to me, and he was going like hell as well. When we had finished, we went into the pub for a couple of pints because we had earned them. Shearing wet sheep was an unhealthy job.

At the end of July I took my first lambs to sell in one of the NFU store lamb sales and they made a good price. I took fifty Suffolk crosses and made £5 2*s*. 6*d*. Trade was better because of the bad winter, as there had been a lot of losses and I was glad of every penny.

CHAPTER 8

Farm fit for a wife

IN AUGUST 1963, at the age of twenty, I decided to move into The Pant on my own. There was an old round table in the garage there, a chair with an arm missing and another with no arms at all. This was my first furniture. I had my old bed from home and I bought an electric kettle, an electric ring for cooking and an electric airer, which was a good buy because nylon drip-dry shirts had become available and that helped me. I also bought an electric fire and a couple of saucepans. I had two spoons, a knife and fork. I also had two milk churns to bring water from the village. This was a bit of a pain because whenever you were in a hurry there would be no water, but I managed OK.

I wanted to be on site because I intended to buy some cows and rear calves.

When I left home my sisters were sixteen and thirteen, so were old enough to look after themselves and they managed alright. I went back to keep an eye on them because Father was hopeless about the house.

That autumn I bought another forty ewes to make my flock up to 200. I didn't do any ploughing that year but got the farm into the Small Farmers Scheme which paid grants on reseeding, ditching and draining. Farmers were given a payment to help improve the farm. We also had a 'headage' payment on the ewes of about twelve shillings per head, that's 60p, so doesn't sound much today but it was a help. Cows' headage was £12, bull calves £9 and heifers £7, so I wanted some cows so I could rear calves and help with the income. I bought my first cow in November; she was a brown Shorthorn with a Hereford bull calf. She had plenty of milk, so I bought a grey Shorthorn bull calf the same day. The cow and calf cost me £76 and the bull calf £6. I had some fun trying to get the cow to take the bull calf but, in the end, he learnt to suck between her hind legs so she couldn't kick him. A couple of weeks later I bought a little Friesian cow with plenty of milk and reared four calves on her. She cost £71 with a bull calf and had given birth to a few calves.

In the autumn I had to change the tractor. When I started using it after the snow went, I found that the frost had caused a crack between the pistons. I had drained the water out before Christmas, before the frost began, but some water must have been left in the engine because after using her for an hour or so she would start to miss and lose power. So I went to Auto Palace in Llandrindod Wells to look for another, and old Jack, the salesman, said they had seven Fergies there and I should go and have a look. I knew one of the mechanics and asked him which was the best tractor? He pointed to one and said she was a good tractor and the only thing wrong was that the tyres were worn on the front and back. I did a deal of £60 for my tractor and £100 for the new tractor, with the tyres swapped from mine to the new one because mine were still good. Old Jack agreed and she was brought out with a mechanic and they changed the tyres and wheels on the yard. That was a very good tractor and went well. It ran on proper TVO fuel – a mix of petrol and paraffin, whereas the one I sold was a petrol conversion, although some people disliked TVO.

I had a lot of hedging booked for the winter. Some of them were big heavy hedges, so Morris Jones said I could have his workman Austin and he moved in with me over that winter. He was so strong and using an axe was easy for him, so he was good to have the other side of a hedge. The only thing was that when you had somebody working for you, you had to keep work in front of them so they earned their pay. We hedged in all weathers and got a lot done. Austin moved back to The Rhos in mid-March ready for lambing.

After lambing, I started on the work for the Small Farmers' Scheme and did some ditching. Lance Lewis from Upper Rhos was out of work at the time, so I asked him to do some of it.

At the May Fair in Builth that year, there were some nice cows and calves. I was looking at these cows when Tom Jones, the auctioneer, came by and I told him that I wouldn't mind a couple of them. He said 'You buy a couple, boy, we will carry them until 9 November.' I thanked him and went on to buy two. I hand-reared ten brown heifers on the bucket and my cattle numbers were growing.

Under the scheme, you could grow rape and turnips for fattening lambs. So I ploughed eleven acres below the road, some of which was pretty steep, but I managed. I put lime and slag on it and sowed it in good time. I had a very good crop with the best rape up to four feet high. I hadn't got enough lambs to eat it, so I let it to Morris and Stan Jones – they also bought what lambs I had left, so it was a good deal. They filled the patch with lambs, who did a good job of eating it all off.

Adjoining the hill was a piece of land about two-and-a-quarter acres that had not been touched because of some old cock-fighting clause in the deeds. The next-door neighbour had the right to cut and take the hay off it, along with another piece of land down the bottom of the meadow, so Ted Breeze had never bothered with those fields. I wanted to put them in the scheme, but the Ministry wouldn't do it with that clause over them. I went to see the neighbour and asked him how much he required to agree to take the clause off – he said he would take it off for £60. I agreed, providing he paid the solicitor's costs, which he did. They sent an old man from the

Ministry to see the field and to measure it, and by the time we had walked up there, he was knackered, so he asked me to step it. I started stepping up the side of the field, but he shouted on me to stop. 'You are stepping too short, get up the top of the field and then step down.' Anyway, it worked out to what we had said it was. When he had finished with me, he went to count cows for Morris Jones. He asked Morris 'Who is that young bugger at the Pant, he's a bit sharp?' I got to know him after that. He walked up from Upper Rhos when the road was blocked with snow once, and I took him in for a cup of tea. When he asked me how many cows I had I told him, but he never wanted to see the cows. I ploughed the piece of land and some adjoining it, and put in turnips. I had a good crop.

That year I had a good harvest again which was good for the next winter. I was glad to get the hay off as I was a bit short of grass because I had ploughed sixteen acres out of my fifty-six. I also turned some ewes and lambs to the hill that joined the top of the farm to ease the pressure.

I was still out working here and there, doing what I could to earn money to buy calves to rear. I always had something on the go.

One night in October, Morris Jones called by as I was feeding my ten brown heifer calves in the orchard. He was looking at them and asked when was I going to sell them. I told him that they were going the next Wednesday in Hereford. He then asked what was I going to make on them and I said £27 a piece. 'Hell boy,' he said, 'you'll have a job. If you make £27 on them, you can sell some for me the week after.' Anyway, Wednesday came and the heifers were in Hereford and they made £27, spot-on. I was pleased – they had cost me £7 each and I had £7 sub on them, so they paid alright. I also took the bull calf off the first heifer. He made £57 and the Shorthorn bull made £35. The two calves paid for the cow, plus some profit and sub of £9 per head.

Farm fit for a wife

The following week I took twelve heifers for Morris Jones to Hereford and made £27 on his as well – he was pleased.

At that time a lady from up above Rhulen called to see me as she had just bought a Land Rover and she wanted me to teach her to drive. She was about sixty and had never driven anything, other than ridden a horse or push-bike. I thought it was going to be a hell of a job but anyway I told her I would go up and give her a go around the fields to start. Fair play, she soon got the feel of it and after a few days I took her out on the road. It was like riding the wall of death for a while, but she did improve, and after a couple of months she passed her test. She wouldn't drive long journeys – she would come down for me to go along with her – but she was the toughest woman I ever met. They farmed a place called Noyadd Rhulen and she could shear, hedge, work horses, plough and mow. Her husband was a wild man as well and once, when the Ministry man came to pass some calves for sub and refused to do it, he chased him with a hook down to his car. He wouldn't go there again.

After the big snow, her husband had a stroke, so she had to do all the farming herself. How they managed over the storm, I don't know. They had stuff delivered to Upper Rhos and went down with the horse and tied bags over the horse's back. The house up at the farm was in a bad state – the top end had fallen in, the front porch was full of straw bales because the door was rotten and had holes in it. They paid the rent for the farm with rabbits. Before mixy (myxomatosis), she sold hundreds of rabbits a year. They would be rabbiting in all weathers, and when they went in at night they took their wet clothes off and put the same clothes on the next morning. She reared a lot of calves and went to Shrewsbury on the train to buy them because they were too dear in Builth. She would bring them back to Builth on the train, then ride home on her bike and return again with the horse and gig to take the calves home. She always had a few calves on the go.

That autumn I went to a farm sale at a farm called Trallwm in Abergwesyn. We had bought stock from there before, and Dai Felix was a friend of my father. When we kept wethers for the wool, we bought them from him, and he would drive them down to Builth and Father would meet him and drive them on home. We kept them until they were five or six years old and then sold them for meat.

Father wanted a cow and calf and Dai had some good Hereford cows. He bought one, and at the end of the cattle sale a heifer came into the ring with a halter on. She had one horn going up and the other pointing down by her face. I said 'If she's cheap, I will buy her' – and I did. She was about eighteen months old and cost me £35. She came home on the same lorry as Father's cow and calf. Many years later that heifer was very important to me. The following Saturday I went to another farm sale and there was a Welsh Black heifer with the same horn problem as the one I'd just bought before, so I bought her for £35 too. At the end of November I had the two dehorned by the vet. They looked different heifers with their horns off and I sold them in the December sale in Builth for £50 each, so they paid well.

That winter I started to do a bit to the house, as I had got engaged to be married, so had to have it liveable by May. The first job was the sitting room where the floor was very poor. Estyn Lloyd, Jack Lloyd's son, came and gave me a hand to do the job. We took the flagstone floor out and dug it down level; we put a damp course in and concreted the floor, and we put new skirting boards all around. It looked much better. We still had no running water, mainly because this job was going to cost a fair bit of money, so I put it off until later in the year.

I worked all hours hedging and fencing that winter, and worked for Jack Lloyd a bit as well when he had a big job on.

CHAPTER 9

Running water, at last!

LAMBING ARRIVED AND went and it was soon May. On 23 May 1964, Margaret and I got married in Clyro church with the reception at Clyro Hall. It went well, although I was still suffering from a hangover from two nights before and when it came time to speak to all the guests, all I could say was 'Thank you very much for everything.' Jeff Davies was my best man, and he made a very good speech, taking the piss out of me as usual!

We took off after the wedding for our honeymoon in Torquay. I had changed my A30 van for a Morris Mini van, which wasn't very comfortable on a long journey. We got to south of Bristol and stayed the night in a hotel. I think the locals knew we were just married because about midnight they were singing 'We'll keep a welcome in the hillside' downstairs. Next day we travelled on to Torquay to our hotel and enjoyed our week looking around the area. When we started for home on the following Saturday, the driving seat collapsed, so we had to stop and buy a biscuit tin to hold it up!

When we got back home and settled in it wasn't easy with only water in the two churns for everything. Margaret went back to work, so I needed another van and I managed to buy an old Bedford box van from a friend of ours in Llandrindod Wells for £27. She was a bit rusty in places but other than that she went well with plenty of room in the back. We soon had small jobs hauling pigs to Hereford for a farmer down the road. This suited me well, as I could also pick up a few calves to rear.

I now had to reseed the eleven acres below the road. I met Jeff Davies one night and told him I had got some walking to do the next day to sow the acres by hand, so he said he would come and help sow it with his tractor and spinner. He came at 8am and had sown it by dinner, which saved me a lot of work. I put a ley of Webbs seed on the patch. It was a good ley and it is still on the field fifty-three years later.

The field came up well but about a month later I got short of grass. There was a nice bit on the field so I let the sheep in to graze it for a few days, then turned them off for a week. I kept doing that for a while and it made a good ley with plenty of clover.

I did my usual stint shearing for Garney Morris and a few contracts of my own. One place I sheared was for Jim Lloyd of Bilmore down by Radnor. He was a bit of a character. I would arrive about 8.30am and Mrs Lloyd would shout for me to go and have a cup of tea before starting. I would come in through the one door into the kitchen and he would come in from the back and say he had been around the cows. She would say, 'Don't listen to him now, he's getting up!' He would soon have some sheep in and we would be away. He would be catching, and his wife would be rolling wool. She would go about 12.15pm and by 12.45pm dinner would be on the table – fair play, she was well organised. He rented Kington Cricket Club field to graze the sheep so he bought some lovely hogs and lambs in Builth fair but had no way of getting them home to shear. I hauled them back in my Bedford van and took them back afterwards, and he was very pleased. Friends came to see him in the middle of the afternoon and he told them not to come bothering him. He'd got Godfrey Bowen there and he was having a job to keep up, which made them laugh.

After shearing, I got Reg Morgan to come and rotovate

the field next to the one I reseeded and did the same with that one too, using Webbs again. They hadn't ever been done before and the leys were poor.

My next project was water. So I ordered 1,100 yards of half-inch polythene pipe and joints for £80. When they arrived, we were ready to go. The first job was to build a catchpit at the spring. It only took eight blocks. I put a concrete base in and assembled the blocks, leaving a hole for the water to go through. The next job was digging the pipe in. We had about 180 yards to dig down a steep path onto the road. I wanted the pipe in about eighteen inches and the first sixty yards was very hard. Again, Jeff came to give me a hand.

We started digging with me on the spade and Jeff on the pick. Once we took the first spadeful off, we were into a stony patch, and it was hard but we got down a fair way. The next day we did get down to better digging. We came to a flat part, about six yards round, and Jeff said that if we could get a tractor up there we could turn around and use a plough with one furrow taken off my two-furrow plough to save some work. So when we went in for dinner that's what we did, but if you saw where we went up and over with my tractor, you would not believe it. At the bottom of the path, there was a nine-foot ditch to get up over and the path was steep. We dug a track for the top wheel and had a go but the front end of the tractor came up, so Jeff told me to put my feet both sides on the front and hands in by the fan belt and lean back – it might be enough to get up. We were lucky and just made it up and over. I stayed on the front all of the way up to the turning area and we dropped the plough in and ploughed down to the road about a foot deep and then across the road, down the bottom side of the road, all the way to our hill fence, saving even more time.

By the next night we had dug down to the road but Jeff said he had another job to do. I met Jeff Langford that night and he agreed to come for a day or two, so we soon got to the hill fence. Again I was lucky. The next morning, I was at the shop in Hundred House and met Kinsley Morris from Llanedw and asked if he had a digger. He said he had a backactor on his David Brown tractor and would dig across the field to the house. He came the next day and dug a trench two feet deep and a foot wide to put a half-inch pipe in. It saved a lot of work by hand. We also dug around the house and down across the yard to put a drinking trough in for watering the cows. Next job was to put the pipe in. I started at the top and joined each 100-yard length. The joints they had sent me were making me a bit worried because the fall of the spring down to the house would be 300 feet and the pressure would be high. The joints didn't look good enough for the job. I decided not to fill in two yards around each joint until we had tested them. A good job we did. When we tested it, they blew one after the other. I took advice and was told to use the same joints as on copper pipe and that worked, but the pressure was too high. When I turned the tap on in the house, the water hit the ceiling and was white like milk! I decided to put a stop-tap in the pipe by the hill fence and put a drinking trough there for the sheep as well – that cured the problem.

What an improvement it was to have running water! It was lovely water from that spring, and it never went dry, even in 1976. We then went on to put in a bathroom, airing cupboard and a new kitchen, so I had a happy wife! We went to Coles of Bilston in Wolverhampton and came back with my Bedford van packed full with a bath, toilet, kitchen units and a food cabinet. We also put in a septic tank, so it was a big improvement.

During that autumn, Jack Lloyd took on a dust round, collecting rubbish – he had a three-and-a-half-ton lorry, which was ideal. So two days a week I was a dustman down around New Radnor, Old Radnor and up to Gladestry. It was OK when it was cold but in warm weather the wasps were a

Running water, at last!

problem. Some of the dustbins were very heavy, so you had to be strong to tip into the lorry, but it must have paid Jack OK.

CHAPTER 10

Struggling to make a living

THE NEXT WINTER I was back on the hedging and fencing in between working for Jack Lloyd. In January we had heavy snow around Hundred House so we couldn't do anything. My wife was working for the council in Llandrindod Wells and her boss asked if I wanted a job hedging down by Clyro on the side of the main road. I went to look at the job and took it on at £1.75 a perk (seven yards). The usual price was about £1 a perk, so it was a good job. I told Jack Lloyd about it and he said he would come as well because he couldn't work, either. They were planting thorn hedges with a fence on the top side, so we hedged separately. I stepped seventy yards on and he started there, while I hedged into the back of him. We did that every day and soon had a long length done as he would do sixty yards in a day. The snow stayed over a fortnight, so we got most of it done and earned good money. We did both sides of the road to Rhydspence pub and two lengths up on the Rhosgoch road out of Clyro.

My cow herd was bigger now. I had got up to eleven cows, making them rear two calves and some on the bucket – it all helped to increase my turnover. On the sheep side, my ewes were getting pretty old so I replaced some and bought oldish ewes to save money. I had some off the top of Glascwm from Jack Morris, Three Wells, who lived just as rough as the Chapmans in Rhulen. We called him Black Jack because he

Struggling to make a living

never washed and lived with bags in the windows. He had no car or tractor and walked down to Glascwm to meet the bus to Builth. The ewes had led a hard life. I couldn't buy them as he would not sell them to me, so Stan Jones bought them for me at £2 each. They were hard Speckle ewes and they did well.

I also bought some better ewes from Mr Davies, Craig-yr-onnen, which were good Speckle ewes – I paid £5 for them. I had done a good job wintering the ewes and I had plenty of hay, so I gave them ample and was looking forward to a good lambing, but on 7 March we had one hell of a snowstorm. I had an Ayrshire cow calve the day before with loads of milk, so I got up early the next morning to get the feeding done and to go to Hereford to buy three calves to go on the cow. It was hellish cold with a hard frost and my Bedford van refused to start, so I borrowed my father's A30 van and went and bought three black bull calves for about £12 each.

On my return home it was starting to snow and blow badly, so I put the calves in the shed and took Father's van back. When I was walking home there was about three to four inches of snow on the ground. I did some feeding and by about 5pm I was worried about Margaret getting back from work. If she came through Hundred House there would be some steep hills. So I took the tractor to meet her. As you can imagine, it was bloody cold with no cab in those days. I kept going all of the way to Hundred House without meeting her and by then I was about to freeze to the steering wheel. So I called in the shop that sold boots, shoes and clothes. I bought a fresh waterproof coat and some gloves, then drove back to the kiosk in Rhulen. There was the van, Margaret had come back the Aberedw way and was in the house with the Lewis family. By then I had three inches of snow up the front of my coat and was frozen. They asked me in and gave me a very large glass of elderberry wine to thaw me out. After warming up by the fire and with the wine, I told Marg we'd better go. So she got in the transport box and sat on a bale and we set off up to The Pant. When we got home she was nearly covered in snow because the drifts were coming out of the hedge, so I had to help her into the house and give her a glass of gin to thaw her out as she was frozen.

I went back out to finish feeding and came in about 8pm for food – it was snowing badly and blowing a gale. We went to bed about 10pm but I woke up at about midnight because I could hear the wind blowing. I dressed and went out to down below the road to check the sheep – they were lying down covered with about a foot of snow and it was covering them fast. I went to the barn and got a bale of hay and I shook it out about twenty yards from where they were lying, then got them all up and they moved to eat the hay. Up above the house, the sheep there were the same and I got them hay as before. I stayed up all night and kept taking hay and moving them around. If I hadn't stayed up, I don't know how many sheep would have died but there would have been a big loss.

With first light the next morning there was a mess. The road was blocked with a big drift and the bad weather had knocked the ewes about. I got feeding and checked the sheep and then went to feed my calves. When I got in to see them they had three inches of snow on their backs and were shivering, so I brushed them down and suckled them on the Ayrshire cow. Once they had filled up, they looked better. I had just finished feeding the cows when Margaret shouted that there was something wrong upstairs as water was dripping out of the lightbulb. I said 'I bet the attic is full of snow!' When I pushed the door to the attic up, there was a foot of snow all over the loft! I got a fire shovel and bucket and cleared it all out or we would have had a mess.

After that, I went back out to look for a ewe and lamb I had put in the pig cot – hopefully to save their lives. The ewe was covered in snow and the lamb was dead so that had been

Struggling to make a living

a waste of time. The snow lasted about a week. The council came with a Hymac to clear our road and tipped the snow on top of the hedge, which caused a lot of work and cost as I had to wire it. It was a cost I could have done without.

During that year there was a general election and Harold Wilson got in as Prime Minister, which did farming no good at all. It was hard enough before, but after that lamb and cattle prices dropped by twenty per cent a head. The Labour Party has never been a friend to farmers and keeps the prices paid to us for food as low as possible.

We had a tough time lambing after the snow, with a few losses of ewes and lambs. I was glad to see May.

At the end of April, Jack Lloyd had a job putting water in at a farm in Radnor Forest and I went to help. It meant putting in a catchpit down in a deep dingle and I went to help pump the water up to a tank on the top so that the water would run down to the house. This meant a lot of handwork. It took about a fortnight but we had water in the house and it was a big improvement for them.

I still did two days a week of work for Jack on the dust lorry, and I was fencing in between to keep the cash flowing. Twice a year I was a postman for a week to let the local postwoman have a week off. She used to do the round on foot, but I took my Bedford van. The trouble was I would meet farmers on my round and they would all want to stop and talk, so it took me as long as it did for her!

I put a field of oats in for the cows the following winter and these came up well. My cows calved lucky, and we had a good hay harvest but this had been a bit of a job. I had 700 baled up and no help, so I got up early the next day and hauled them day and night. It was a lot of work unloading into the bay and then putting them in their place, but I was fit in those days.

Again, I had a good season shearing with Garney Morris and my own contracts as well, though my back was starting to play me up. I suppose all the physical work I had done since I was fourteen was catching up with me.

Selling lambs that year was hard work. I had used a Border Leicester ram on some Welsh ewes and managed to sell the ewe lambs OK. The wether lambs were harder to sell, so I took eighteen in the back of my van to Kington to sell as store after the ewe sale. The trade was hopeless, but I managed to sell them afterwards for £3.50 and had to deliver them on my way home to Gladestry. Morris and Stan Jones had come down with me as they were selling some Kerry yearlings. After the sale they suggested going for a pint and I said OK. We had about two pints each and loaded the lambs up and headed for Gladestry to deliver the lambs. Morris knew how to find the farm and we got them there, unloaded the lambs and headed for home. Their van was at Boxes Bridge below Hundred House, so I dropped them off. Stan was first to get out and immediately staggered backwards across the road into the hedge. Morris got out and was on his knees laughing at Stan – he was drunk on two pints of beer, and he was driving from there!

I wanted to get more land so I asked my father if I could have part of his farm but he refused, although he wasn't doing much of a job farming. There was no land around coming up for sale for years, so I had no chance to expand.

I carried on working for Jack Lloyd through the autumn, and when November came I had orders for hedging again. I did a lot of hedging for the farmer who lent me the money for The Pant, which reduced my interest bill. I also did a hedge for John Lloyd near Gladestry and then for Tom Davies, Llanion, Hundred House. He was a champion hedger himself at twenty-one, and I enjoyed hedging for him because he knew what he was talking about and would come by two or three times a day to offer a bit of advice.

It was a better winter for working, but by spring my back was playing me up again. I couldn't work as hard as I wanted

Struggling to make a living

to. Anyway, lambing came and we had a fair lambing, and the cows were doing very well.

I decided to go and see my bank manager to see if I could borrow enough money to build a shed at the back of the barn. It was a level site and would have been easy. I only wanted enough money to buy the blocks, timber and zinc, as I would build it myself. He refused point-blank to lend me a penny.

I lost my temper and told him if he would not lend me the money, I would go over to the auctioneer across the road and put the lot up for sale, as I needed the income from rearing more calves from that shed. We now had a baby daughter, Amanda, so Margaret wouldn't be working for a while, which was all exciting but a bit stressful!

He told me to do what I liked, he wasn't going to help me. So I marched across the road and told Tom Jones about it but he said 'Don't do it, boy, you stick to, it will get better.'

News got around that I was going to sell and about two weeks later my cousin came by to ask if it was right. I said it was. He asked how much I wanted because he was getting married and needed a place to live. I asked a fair price and he said 'I'll have it.' I told him I would need time to sell my stuff and get a job. I wanted a job in the Cotswolds, as Jeff Davies had taken a job down there and was very happy.

63

CHAPTER 11

Pastures new

THE FOLLOWING WEEK there was a job going as a shepherd near Bibury on an arable farm. I rang the number and talked to the owner who said he was coming down to Sennybridge to a bull sale and would call back by on his way home at about 4.30pm. On the Friday, this huge Mercedes car turned into the yard. It was the biggest car I had ever seen. When Gregory Phillips got out, I noticed he only had one arm but he managed driving no problem. We had a long chat and he said that I should come down to Bibury Farm and look at the house, farm and the sheep. We went down on the Sunday, and he met us. The house we would live in was half the farmhouse. It was a big house so we would have plenty of room. Margaret liked it and there was a good garden too. The farm itself was 1,100 acres, mostly corn, but sheep grazed the land they couldn't plough. They were Scots Half-bred ewes cross with a Suffolk tup. I met the shepherd who was retiring. He was in his seventies and told me Mr Phillips was a good boss. I was to be paid £21 per week and have a free house plus £2-a-week dog allowance.

I took the job and said we hoped to be there in a month. I carried on selling the stock and tackle and managed to sell the cattle without having to take them to market. The farmer I had borrowed money from bought three cows, so that helped. Owen Mills, who bought the farm from me, bought some of the sheep, the tractor and plough. I managed to move it all within the month. As we would be travelling back and forth to Clyro to see Margaret's family, we bought a Ford Cortina car for a bit more comfort, especially as we had the baby to think of as well.

It was about the end of August when we loaded our stuff up on the removal lorry and said goodbye to Rhulen. As we passed my home, my father and sister were at the gate at the bottom of the lane. I said we were off and that's when he said that I could have the farm. I told him, 'You had your chance, we're off.'

We arrived in Bibury and unloaded our stuff and soon settled in. On one of the first days at work, I was in the barn at 7.30am when the manager came in to give the daily orders to the men. There were seven men working on the farm as tractor drivers, stockmen and maintenance men, but mostly tractor drivers to handle the arable side. The manager was a nice gentleman and, as the shepherd, I was never given orders. It was up to me to ask for anything to do with the sheep.

The main job with the sheep was moving the electric fence to the next field where you intended to graze them, because some of the walls were not maintained. After what I had been doing, this was very easy and managing the flock of 430 ewes and 120 lambs was no problem. I had two good dogs so I could manage most jobs OK. If I wanted help there was a chap who would come and help. He was pretty stiff – he had been injured in the war – but it was better than no-one. The boss, Mr Phillips, always wanted to pick the lambs to sell, so that meant an early morning. He would come about 6.30am. I usually had them in ready, and he was good at picking the best to sell. I marked them and we raced them out afterwards.

We grazed a lot of deep valleys that went through the middle of the farm. Very often in the summer, I would be checking the sheep and Mr Phillips would be sat in the middle of the sheep in his Merc – he loved his sheep. When we got to mid-July, they would start combining the winter barley and they had three, twelve-feet cut combines. It was quite a job to haul the corn to the dryer and the manager asked me if I would help. I agreed, so he found me a tractor and trailer and away I went. I watched the ones in front of me picking up from the combine – you had to get into the right gear and pull under the combine. When that one was empty, you moved onto the next and if you had room in the trailer, the next one as well. We would be four or five tractors and trailers and a four-tonne lorry going flat out all day. It was a busy job. I enjoyed it and we kept going until the first week in September.

The farm which was near Cirencester was a dairy farm, the rest mostly arable. Adjoining Bibury, there were two 1,100-acre farms, both with managers who were the brothers of our manager. My boss inherited all of this and it was farmed well as he oversaw all of it. Some of the farms had pheasant shoots and wild partridge as well. I used to go beating pheasants on a Saturday for a change.

Through the winter I had to graze kale and swedes with the sheep. As I was not used to sheep this size, they were big ewes, I took advice from the manager as to how much feed they needed.

We started lambing on 20 February. They would build a straw bale pen, putting about five bales high around the outside, leaving a gateway in and out. Then they put a roof, seven feet out, into the pen, using poles that had been used for years. They put ash forks on the bales and long poles up to the ash forks along the front to hold the ash poles straight and put sheep netting over the top to keep it firm. Wooden hurdles were put around under the roof to keep the lambs dry and the middle of the pen was in the open air but well shaded. This was how it had been done for years. The site was moved every year to stop any build-up of parasites or infection, and they had a shepherd hut parked by the shed with a fire in it to keep the shepherd warm on a cold day.

Two weeks into lambing, I started to have a few with twin lamb disease and lost them. I upped the feed, and they were OK. It was a fairly hard lambing because the ewes were not in a good enough condition for the number of lambs they were carrying, but we still reared 187 per cent with a lot of work. I learnt a lot that spring and knew there could have more lambs alive with better feeding. The hogs did OK with 117 per cent, with ninety out of 120 having lambs.

When it came to shearing at the end of May, the manager told me he had two men who usually helped with the shearing. When I told one of them, old Bill Tibbles, about the shearing machine, he said he would get it going. It was a Lister oil engine with four heads, one off each corner with a tank on top of the engine filled with water. It was a very heavy old thing on iron wheels, but they said it went well. We got started the next day and I had to learn the Bowen-style to shear these big sheep, so I bought the book *Wool Away* where it explained the style.

When we started shearing Bill Tibbles was on one corner, Phillip Legge on the other and me on another. The old engine started up and ran the machine well. They started shearing their own style – they sat the ewe down and sheared one side with the right hand, then without moving the ewe, sheared the other with the left hand, which on big ewes was less hard on the back – but I did notice they were cutting the sheep a lot. When we stopped for coffee, I asked if I could look at their hand-pieces. I noticed they had never been shown how to put a comb and cutter on – they just put the comb right to the back and put the cutter on, which meant the cutter was out above the comb and bound to cut the sheep. I told them I could adjust their combs, and when we started again, what a difference! They didn't cut the sheep and were pleased I had told them. Old Bill said 'I wish some bugger would have told us that years ago.'

I was struggling a bit with the Bowen style – I had been shearing on the bench for ten years – but soon got the hang of it. We took three days to shear my sheep. We had moved to different lots of sheep several times across the farm and it took time.

By the end of May we started selling lambs and I was surprised how heavy they were by then, over forty kilos, and they hadn't had any creep feed, only grass. They had grazed some young leys before shutting off the hay, after this they grazed the valleys through the summer. When we weaned the lambs they went on to lattermath and kept growing well. The boss told me to pick fifteen hog lambs for the Christmas show in Cirencester – they had to be ninety pounds and under or over ninety pounds to fit the classes. These had to be kept separate and once we got to November, they were allowed some dry food.

That year my youngest sister got married and her new husband took over Llwyn Tudor from Father.

On 23 October we had a baby boy, Michael, who was born at home with only me and the midwife present. All went well and Margaret's mother came on the bus to stay for a week to help, so we now had a daughter and son.

Father had a stroke when he was sixty-two years old and passed away seven years later. He never remarried.

I have two younger sisters, Vera who is the older, and Janet who is a few years younger and we regularly keep in touch, reminiscing over lunch.

When my mother remarried she had three further children, one boy and two girls. Unfortunately we do not have regular contact.

About that time, Jeff Davies left his job near Cirencester and bought a farm, The Pant, Merthyr Cynog, and he moved back to Wales. The Pant was a dairy farm and a good run for sheep.

In November I had a go in a hedging match near

Gloucester and didn't win, but enjoyed the day and met a great character, Arthur Vaughan. I was hedging crop and pleach style and I could see he was doing Breconshire style. After the match finished, I went and had a word with him. He said he was from Breconshire but was a shepherd for the Price family at Exton north of Gloucester. He won third prize and I had fourth in the match. First and second were Cotswold hedgers. A father, son and a daughter hedged as well, but their hedging wouldn't do in Wales as they hedged 'very thin'. After our initial meeting, Arthur and I saw a bit of one another around the market and became friends.

Two weeks later I went to a match in Black Bourton and decided to do their style, but better. I came third behind a father and a son. I thought I had beaten them, but it was a local judge, which makes a difference. The father, Mr Padget, knew I had done a good job, and when I was packing away my stuff after the prize-giving he filled the cup and came over to me and said 'Have a drink.' He told me that I had hedged 'bloody well', so I was pleased he knew I was a 'fair hedger'.

By the time we got to December 1967, foot and mouth disease broke out and livestock sales and markets were closed, meaning no Christmas shows. My lambs went straight to the slaughterhouse.

That winter, I joined the local darts team in the Catherine Wheel pub and played on Wednesday nights. I met a lot of locals that way. There was a gardener and a blacksmith in the team and they were good fun. My drink was Glucose Stout, which I liked, and we travelled to pubs around the villages in the area. I had the third highest points in the team that winter.

My sister Vera came to stay for a week. She loved the Cotswolds and decided she wanted to find a job there. On the Saturday, I picked up the *Wilts and Gloucestershire Standard* newspaper and there was a job going as a dairymaid near

Winchcombe, so she asked me to ring and see about it. I arranged to take her on the Sunday afternoon. We arrived at the farm on the edge of town and met a very nice lady who seemed to be the boss. She showed us around and asked Vera about herself. After an hour, she offered Vera the job and she took it. She moved down and was soon milking seventy cows twice a day, and enjoyed every minute. There was a son on the farm about the same age and very soon romance started. They later married and had three daughters.

During the winter, I did some hedging on the farm because the two fellows doing it were not very fast. They would only do about twenty-two yards a day. The old chap fed a lot of cattle on his way to hedge, while his son would bike down to his hedge by daylight, sharpen his axe and hacker, then have a cup of tea before starting. If they came to a three-yard gap in the hedge, they stopped and started again after the gap. Bill Tibbles would put rails across the holes in the hedges every gap after they had finished. They hedged too thin to keep stock off and gave me the hedges they didn't like, but I could manage them OK.

That winter, I did a better job of the ewes and fed them a lot of good clover hay with kale and swedes. When we came to lamb, I fed plenty of cake, but I lost a couple of ewes because they were too fat, so I eased back on the cake and they were alright.

Lambing was good with a lot of twins but too many triplets. I adopted a lot onto singles but they could not keep up, so I bought two buckets with twelve teats and started putting what we couldn't adopt on the milk powder. Very soon, I had a lot and the kids from some houses down the road started coming up before going to school to feed the lambs. In the end we reared sixty-seven tiddlings. It was a lot of work, but we had to do something.

During lambing a TV company called by, wanting to make an advert for Wonderloaf bread. They wanted a lamb for a little girl to hold and show the wonder of a new lamb. I agreed and they came the next day with a girl of about six. She was as white as a sheet and had never seen a lamb before. It took a quarter of an hour to get her to hold the lamb, but we made it in the end. They also wanted to fly a kite on the field, and once this was done they left. This was my first time doing anything for the telly.

When we marked the lambs, we had reared 854 lambs from 425 ewes, so the boss was pleased. The hogs reared about 120 per cent, which was good enough, everything was doing well.

On 13 May we sheared the hogs as they were getting stuck on their backs. We were still giving the hogs a bit of rolled barley to keep the lambs going.

We carried on and sheared the ewes because they were fit enough. We started selling lambs by the end of May and had plenty ready by then. It was a good season and we cut the hay at the beginning of June. One Saturday morning that month, the manager came into the barn (we worked until 10.30am on Saturdays). They had baled 3,000 bales of good hay on the Friday and he asked if we would work all day to get them in. Three people refused. I was walking with him across the yard and I told him they were lucky they were not working in Wales, or they wouldn't have a job on Monday morning. He replied, 'Oh well, that's the way they are.' I said I'd work with the rest of the boys. We started and cleared the hay by 9pm that night. It poured with rain on the Sunday, so the manager was pleased.

In mid-July, the boss rang one Sunday night and asked how many lambs were ready. I told him there were a good many and he said 'There's a bit of trade going. Will you have them in at 5.30am and we'll take what are ready?' I had them ready in time and we pulled eighty good lambs out and away they went to Banbury in the lorry. The manager and I went to see them. The old market only had small pens, so eighty needed eight pens. When they came to sell ours, which were forty-two kilos, the first pen made about £10.50 and the auctioneer asked the buyer how many pens he wanted. He took the lot, so that was that. This was the only time we went to Banbury, we usually went to Cirencester.

We weaned the rest at the beginning of August and the boss gave the same orders as the year before to pull out thirty lambs for the Christmas show. I was hauling corn again off the combines, and if any fencing was to be done with the sheep, old Walt Williams did that because he wasn't much good on a tractor. I looked at the sheep at 6.30am and was hauling corn after 8.30am until 9pm.

We went to see Margaret's family every six weeks or so. They lived above Clyro by then. We changed our Cortina for a newer one, but it was no better than the old one, just a bit more modern and a new shape.

As already noted, one Friday night, I had a phone call from my cousin telling me my grandmother was in Hereford hospital. So I said to Marg, 'I'll go to see her tomorrow.' I set off at about 11am, arrived there and a nurse showed me where to go. When I went into the small room there was a lady at the other side of the bed and, after about quarter of an hour, I realised this was my mother. I hadn't seen her since she left home. On the way out of the hospital, she said that Gran could not go back to her flat in Gladestry, she would have to come and live with her on the farm between Rhayader and Llangurig, so that is what happened. It was a bit of an ordeal meeting the man my mother had run off with, but I had to bite my tongue in order to see my grandmother. She had always been very good to me. After a while, mother and I got on OK and we were friends until she died aged ninety-five.

In the autumn, my boss asked me what we were going to do with the sixty-seven tiddlings we had reared – they had been fed all the summer but wouldn't fatten. He said the best thing was to sell them store and we should take them to Honeybourne fair, which was coming up the following week; also to pick fifty nice Suffolk cross ewe lambs as well. We took them and there was a prize for the best pen, which we won as well as having top price of the day. The tiddlings made £8.50, but at least they were gone and we did not have to fatten them.

My Christmas lambs were doing well, so when the show arrived we took them to Cirencester market. The first pen of heavyweights weighed 140 pounds each, the second pen about 135 pounds each. When the first pen of lightweights went on the scales, they weighed ninety-and-a-half pounds. The man on the scales said to let them go because that was the rule – ninety pounds and under, or over ninety pounds. They graded the first pen at sixty-nine pounds in weight and the lightweights at forty-five pounds. We won first prize with the heavyweights and second prize with the second pen. We also won first prize with the lightweights, so we had a good day. There was £27 in prize money, but I never saw a penny! The manager did buy me half a pint of beer in the pub afterwards. It was a good end to the year.

CHAPTER 12

Back to Wales

I DID ONE more lambing there and decided to move on because my weight was going up every year. I was now sixteen-and-a-half stone. I had been eleven-and-a-half stone when I went there and it was a healthy place to live, but I was looking for a new challenge.

I saw a job in Wiltshire with a Lt Col Houghton Brown. He had won the *Farmers Weekly* lambing competition with a lambing percentage of 200. We went down for an interview, but the house was poor and the wages less, so we came away. He did ring me about a week later offering more money, but Margaret did not like the house, so I said no.

I then saw an advert for a Shepherd/Cowman up in north Wales on the Rhug Estate. So I went up for an interview and met Lord Newborough, the owner of the estate. He was a bit of a character and wanted me to start as soon as possible. I informed him I had to give one month's notice.

The job was to look after 1,040 ewes and 350 Blue Grey cows from May until November. All of the lower ground was for growing corn, so the stock was up on the high ground, of which there was plenty – 2,200 acres in all. We were to live in a newish bungalow up on the top, overlooking Corwen and facing the Berwyn Mountains.

I took the job and we moved at the end of May, and I started getting to know my way around the estate. The cows were run in bunches of seventy, with one Hereford bull. The sheep grazed the banks and consisted of 600 Welsh Half-breds and 440 Beulah Speckle ewes. Lord Newborough asked me what percentage these ewes ought to rear. I told him the Speckles, under good management, should rear 150 per cent and the Half-breds 160 per cent. I was sticking my neck out a bit, but they were good ewes, mostly three and four year olds. When I asked what the lambing percentage had been the previous year, he said 111 per cent. I thought 'I can do better than that!' At that time he didn't have a manager, but had appointed a man from Scotland who couldn't start for a month, so he was in charge himself. As you can imagine, 350 cows would clear a lot of grass and they had to be moved around pretty often. I was allowed to have a man from the estate to help if I needed, so we managed alright. One of the bulls was a bit nasty and would come looking for you, so I usually had help with him. I had a good dog as well. The bull would even stand the Land Rover; he was the best bull out of the seven. I'm sure he was about a ton in weight.

The new manager, Bob McClain, arrived and he seemed OK. Soon after shearing started and I said I was happy to shear the sheep with a bit of help. When we sheared our own flock, the manager said we had to go to Caernarfon where Lord Newborough owned a place called Belan Fort, which was an old airport in the war. The shepherd there had left, so we loaded up on the Saturday morning and about five of us went there and set up in an old building. I went to gather the sheep, which were all Welsh ewes with Wiltshire Horn cross lambs. They looked well and I got them in a bunch and fetched them into the building. Bob McClain had a good dog as well, so we soon had them in. After a cup of tea we started shearing. Bob could certainly shear; he had been taught by Godfrey Bowen, so had a good style and took off at a fair pace. I stuck to him, so we were clearing a few sheep and when I looked across at about 11am, he was steaming through his string vest. I thought 'I am warming you up now, mister!' Anyway, we stopped for dinner about quarter-to-one and he went out and drank a full bottle of lemonade straight down. When we had dinner, he said he couldn't shear any more; he felt ill and had a headache.

I asked a young chap of about nineteen, Gary Jones, if he could shear and he said yes. He could do a few and Bob was OK with it. Gary was pretty good, so we sheared that lot. Lord Newborough then said there was another lot down by the village, so we went down there and got stuck into them. By now it was hellish hot and about halfway through we stopped for a blow. Opposite where we were shearing there was a beach. I said to Gary I was going into the sea to cool off and he came as well. It was lovely, then we went back to finish shearing. We did another small lot from over the road and finished about 6pm. Lord Newborough asked us in for some supper before we went home and gave the manager £20 for a drink on the way home. Belan Fort was on the edge of the Menai Strait – mostly underground with walls eight feet thick. After supper, I said 'Let's go', as it was cold down there after sweating all day. We came back to the Goat Inn just outside Corwen and had a couple of pints.

During the summer a Corwen farmer rang me up and asked me to go and see a dog with him near Wrexham. He wanted a cattle dog and this owner was retiring and wanted to sell his dogs. We found the place and he was a nice old boy. He took us to show his dog. He was a big, rough-coated dog and worked OK on sheep and the chap assured us he worked well with cattle too. I noticed a black smooth-coated bitch that came with us down to the field. I liked the look of her but she was inclined to creep around behind you as if she wanted to bite you. The chap I'd come with bought the dog for £15. I asked about the bitch too and whether she bit. He said she had never done any harm. I offered him £13 and he replied that, as we had bought the other one, that was alright.

Back to Wales

The chap I was with had a Morris 1000 van so we put the two in the back, but when we got back to his farm, we couldn't touch either of the dogs, they would just bite us! We drove the van into the barn, shut the door and let them out. I managed to get a chain on mine and so did he. I took mine home and tied her in a Nissen hut. Next morning, I went to see her – she was called Fly – and she was on the end of the chain with her teeth bared. I thought 'What have I bought!' After lunch, I went back to see her again and she was curled up asleep, so I took a gamble and put my hand on her head and sat down by her, talking to her. We became friends, and when I fed her that night she was pleased to see me. In fact, she was one of the best dogs I ever had. She would do anything with cattle or sheep, but you had to watch her with strangers.

I never had any trouble with the mad bull either. Fly would shift him, no problem, and moving cows was easy.

That autumn Bob McClain said that we needed to buy about forty Speckle ewes and asked where the best place was for this? I suggested Builth Wells, so we came down to the sale. We walked around and he picked a pen of yearlings from Llwyngwilym, Beulah. His judgement was right. When the judging had been completed, they had a red card. I told him. 'They will cost you a bit now,' but he said, 'We are having them,' and he told me to pick another pen. I picked a pen of the Weale brothers out by Hundred House. We bought the ones he had picked and Tom Thomas gave some luck and the first prize card. We bought the ones from Weale brothers for a lot less.

The following Thursday, we came down to the ram sale and bought three rams. We then went to Llangollen and bought forty Half-Breds. They were popular then.

The third week in September, the All Wales Ploughing and Hedging Match was held at Trefnant Farm, Denbigh, and I decided to go. I asked Bob McClain for the Saturday off. He asked where I was going and said he would come and have a look – he had never seen hedging.

It was a foggy morning the day of the match. I managed to find the farm and the hedge, which was planted thorn. I found my length and got stuck in. When the fog lifted it was a lovely day. I was getting on well, and by 1pm I had my piece cut down. I had a cup of tea and some food, then carried on topping the hedge off. Bob came over after a while, and he said he had seen all the hedges and he thought mine was the best! He asked me whether I was thirsty and came back with two pint-bottles of beer. I drank one and carried on. As I was getting near to finishing, Percy Goodwin, who was judging, in his Radnor dialect said 'How bist thee, boy? Ana seen you for a while, you done a good job here.' That raised my hopes and I worked hard to finish as best I could. At the prize giving I was called up first and was overall champion, my first Welsh Championship. I've won nine more since then!

During the summer Bob had been buying some big Welsh Black bullocks from Dolgellau to sell, as he had the idea of having a sale on the estate. These bullocks had large horns, so he decided to dehorn them. I told him not to do it too early but he did it mid-October. They were grazing some fields next to where I had lambs, so I was going down there every couple of days to check on them. One day, I was going up near the fence joining these cattle and there was a fallen tree. Three of them were rubbing their heads in the tree and I could see they had maggots in their horns and were losing condition fast. I went down to the office and told Bob, but he wouldn't believe me until I told him I had seen the problem before. They got them in the next morning and had to have the vet to treat most of them.

The sale was organised for the end of October. We made some pens to hold about 520 cattle, which included 350 weaned calves from the Blue Grey cows, sixty Welsh Black bullocks, sixty aged Blue Grey cows, and fifty Black Hereford

Back to Wales

cross weaned calves reared down the bottom by a Canadian chap. His name was Green and he had reared about 120 calves in all on nine Friesian cows. Sale morning was a hell of a job, sorting the weaned calves into lots, penning them up and putting numbers on them. The cows were sold in fives, so it was not as hard to sort them. The Welsh Blacks took a bit of handling, they were big bullocks. The auctioneer had drovers, so we only had to oversee that they went into the correct pens. That night, after the sale, Bob said we should go to the Owain Glyndŵr pub in Corwen because we had earned it.

We then had seventy in-calf heifers down from Scotland. When they arrived they were very wild but soon settled down. A couple of weeks later they started slipping their calves, so it was my job to adopt a calf on them. I asked Bob where to get one, he said from Mr Green. I had four from him that week. The next one was a lovely Red Galloway x Shorthorn and she was totally mad. So I drove the Land Rover by the side of the dead calf and put a long piece of string on its leg and then got back in the Land Rover. I pulled this calf back to the yard with the heifer following, opened the shed door and threw the string out through the muck-hole. Then I went round and pulled the calf slowly in and the heifer followed. I ran round and shut the door just in time and got him skinned. Next, I went down to Mr Green and picked a good Black Bull calf, went back to put the skin on the calf and shoved him into the shed, opening the door just enough. The next morning I went to look, and he was sucking away, so I breathed a huge sigh of relief. Two more died and it was another week after that before they started calving well and we didn't have any more trouble. The problem turned out to be brucellosis, and skinning the calves gave me the disease too. They were culled at the next test. When I have blood tests now it still shows up.

The cattle went down the bottom to some parkland in the

middle of November, so I only had the sheep to winter. We also kept 200 Speckle ewe lambs as replacements for next year. I did a bit of hedging in the winter, including some for the farmer who asked me to go and see the dog.

On 11 November we went to Liverpool to see some friends. It was fine in the morning but by about 3.30pm it started snowing there, so I said to Marg, 'We had better go, if it's snowing here what's it doing in Corwen?'

We started home and the snow was getting deeper every mile between Mold and Ruthin, so it was dragging the bottom of the car. We got to Corwen, and halfway up to the bungalow we came across some oak trees that had lost big boughs on the road. I walked back down to our next-door farmer who came with a light and a saw and we cleared them off the road. I thanked him and we managed to get home. I went in and changed and went around the cows as some were calving. There was about nine inches of snow down – that was the first of twenty-seven times the Berwyn Mountains were white that winter.

When the cows wintered down on the parks, they were fed straw and loose beet pulp. Lord Newborough had another farm near Wrexham where he grew a lot of sugar beet and had pulp carried to the Rhug. This was alright for the cows, but the heifers were not doing well enough to rear a calf and they were losing them. We were putting the heifers in and feeding them well to get some milk back, then putting a Black calf off Mr Green on them. In the end they were about forty down. I also had to feed loose beet pulp to the sheep, so we had a lot of wooden troughs made at the sawmill in Rhug. As we got nearer to lambing, I suggested to Bob they might need better feed, so they bought a load of coarse mixture and mixed it in half-and-half which made it better. The ewes looked heavy, and with no scanning in those days you had to judge how much to feed.

One Saturday morning I was down at another farm

FARM NEWS, JULY 15, 1967 13

Shearing gang sets a new pattern in the Mid Wales hills

A NEIGHBOURLY good turn that snowballed as it went led 47-year-old Mr. "Garney" Norris to set up an enterprise that has changed the character and tempo of the communal shearing that have for generations been a traditional feature of the hill sheep farms of Mid-Wales.

Now a dozen men shear a flock that formerly called for the mustering of as many as 80 — and they do it in less time. "This is the best thing that ever happened for the hill sheep farmer," said 75-year-old Mr John Jones above the clatter of the machines at his Dolgoch farm, in the hills near Tregaron, Cardiganshire.

And it happened because Mr Morris, a farmer's son from Cwmforest, Talgarth, helped a neighbour with his shearing on a Saturday afternoon after he had come out of the Services. "Then I took a friend with me and soon we were taking a few days off for shearing", he said.

Desperate

"I started contracting farmers with four men. Year by year the demand has grown and now we are shearing 50,000 sheep in a season." Mr Morris now lives at Llanyre, Radnorshire, and finds almost all his shearers from within the county.

Said he: "I think this all came about because the standard of hand shearing had declined and farmers were getting desperate over the job of getting their sheep shorn. When I was young a good man would shear up to 100 sheep a day by hand. In recent years the number had shrunk as low as 30".

Confirmed Mr Jones. "Even at that, there were a lot of poor shearers. When we were hard shearing, there were always sheep dying afterwards. Now very few die after the contract shearing. I don't know what we would do if we failed to get the job done by contract".

The shearing gang starts on the job at the end of May with lowland sheep and push up the hills as the season wears on, finishing in mid-July. "The hill-billy men were auspicious of our work to begin with", Mr Morris said. "They thought the machines would kill the sheep.

"But first one took the plunge and the others watched. Then each year somebody would ask us to do theirs next year. I've got some pretty good hands — they come from Bleddfa, Pantydwr, Aberedw, Hundred House, Llangunllo, Llanbister, and all over the county.

Like a bomb!

new man, he is likely to improve in his first week so that he shears 50 more sheep on Friday than he did on Monday".

The men all shear Bowen-style. They have a five minute break in each hour, a half hour for dinner and another for tea. The shearing day is from nine o'clock to about 5.30 p.m. but before and after the gang may well have done a 50 mile trip.

They shear only. All the ancillary jobs are still done by the traditional local co-operation — gathering, medicating, catching, wool strapping — and in that sense the event is still a gathering of neighbours in the hills.

A catcher (right) waits with another sheep so that the shearer is not held up.

Newspaper cutting with the news that Garney Morris had set up the largest shearing gang in mid Wales (late 1950s)

Ford Cortina parked outside The Pant, Rhulen, purchased to travel back to Wales from the Cotswolds (1966)

Tom and Margaret before they were married at her home in Clyro (1961)

The first cows and calves purchased by Tom at The Pant, 1961

Tom and Margaret's wedding, 23 May 1964, at Clyro Church

Lambing pen built out of straw bales in the Cotswolds (1967)

Ty'n y Cwm hill (1980s)

Sheep making their way to the hill above Abergwesyn from Ty'n y Cwm (1980s)

Silage pit built in oak wood at Aberannell (1981)

Lambing shed set up at Ty'n y Cwm, built in 1982

Snow-covered Ty'n y Cwm, looking down over Abergwesyn valley and the surrounding area in 1982

Beulah Speckled Face ewes, indoor lambing in 1983

Anthony Andrews, star of Play for Today's *Z for Zachariah*, filmed at Ty'n y Cwm in 1984

Margaret and the family dog, border terrier Rip, a bit of a character (1980s)

Tom commentating at a YFC shearing event (1980s)

Tom's trophy for overall British Champion Hedge Layer, 1986, in Ledbury, out of 100 hedge layers

Tom showing off his trophies, with a wild hair-do at home at Aberannell, 1986

Example of a crop and pleach hedge Tom did while competing near Leicester in 1988, missing out on the championship again by half a point

Builth Hedging Match held in Newbridge-on-Wye; Tom was overall champion on the day (late 1980s)

Tom competing in the veteran shearing at the Royal Welsh Show (2005)

Tom and his four grandchildren, Greg, Paul, Kelly and Amy, when they were younger (2009)

Tom speaking at a YFC event when president of the Brecknock Federation (2011)

Tom with other Brecknock Federation officials (2011)

Tom had the honour of showing Her Majesty the Queen around a display of Welsh livestock at Dolau in 2002 to celebrate her Golden Jubilee

Ray Davies, Dai Fairclough and Tom in the Flinders Range, Australia (2012)

Wool handling shed in Australia (2012)

Merino sheep all ready for shearing in Australia (2012)

Venue of small shearing in New Zealand where Tom commentated the week before the Golden Shears in 2012

Tom was team manager of Wales for the Golden Shears in New Zealand in 2012

Shearing shed, eight shearers on the floor in New Zealand, during 2012 visit

Bales of graded shorn fleeces waiting to go off to the grading station, New Zealand

Tom had the pleasure of welcoming Princess Anne to the Shearing Centre at the Royal Welsh Show in 2012

Tom's 70th Birthday (2013), L–R: Julie, Greg, Mike, Paul, Amanda, Phil, Kelly, Amy, with Tom and Margaret sitting

Thursday, July 30, 2015

4

Express News

PM honours showground shining light

PRIME Minister David Cameron visited the Royal Welsh Show - and presented a special award to one of its stalwarts.

Tom Evans, who has been a steward at the show and commentator at the shearing centre for more than 30 years, was recognised with one of the PM's points of light awards given to recognise outstanding individual volunteers.

As well as his duties at the shearing centre the 72-year-old is the president of Beulah YFC and has also served as president of the county Brecknock YFC.

He has devoted countless hours to passing on rural skills such as shearing and hedge laying to YFC members and also worked with the former Agricultural Training Board.

The sheep farmer from Beulah said: "I was very proud and honoured to meet the Prime Minister, it's not very often you get awarded things like that."

Tom was presented with a certificate from the PM outside the shearing centre when the Conservative leader visited on the final day of this year's show, Thursday, July 23.

The PM, was presented with a singlet when he visited the shearing shed at last year's show, but Tom said it was his co-commentator Steve Meredith who raised a smile from Mr Cameron when he quipped over the PA system "it is something to wear around the garden on holidays".

"Steve is our resident comedian," said Tom who said he hadn't asked the VIP visitor whether he'd worn the vest.

Tom started shearing when he was 14 and bought his own farm when he was only 19. Since then he has dedicated his life to farming and to passing on traditional farming knowledge, such as hedge laying, to a new generation through the Young Farmers' Club and the Royal Welsh Show.

He also learnt hedge-laying when he was 13 and won the All-Wales Hedging Championship 10 times. A highlight of his 54 year career was to beat 100 competitors to win the British hedging title.

The visit from the PM had capped a successful week in the shearing shed, said Tom with a large crowd enjoying the action on Wednesday with New Zealand and France in international tests.

"It was the biggest crowd we've had since we held the world championships about four years ago."

Tom said former world champion David Fagan, from New Zealand, retired at the show where his son, Jack, won the open championship.

Tom received Prime Minister David Cameron's 'Shining Light Award' at the Royal Welsh Show in 2015 for his services to Young Farmers' Clubs and the local community and Royal Welsh Show

Sheep patiently waiting to be shorn at Mount Linten in 2017, during Tom's last visit to NZ

Tom has a very full shelf of trophies showing his hedging successes over a period of 61 years

Tom and a shepherd's crook – a gift for his 80th birthday (2023) from one of his grandchildren, made by Colin Davies, Church House, Aberedw

Ruth Rees Photography

A rare photo of Tom and his sisters, Janet and Vera, taken at his recent 80th birthday party

Ruth Rees Photography

Tom still checking stock at the age of 80 on his quad bike!

Tom received his MBE from the then Prince Charles in February 2020 at Buckingham Palace

building when Lord Newborough came on the scene. We had a chat and he asked if there was anything I wanted. I jokingly said 'A big shed to lamb the ewes in.' He asked where I would put it and I told him there was a good site at the back of the farm buildings. He had a look and agreed. He would see what he could do. Lambing was approaching fast, so to be any good it needed to be up quickly. Within a week, he got a digger on site and the following week a huge building went up. One night, when the frame was up I looked and I wasn't happy – some posts along the front were out of line by a foot or more. Just as I was leaving, Lord Newborough arrived and asked if it was OK. I said it wasn't and asked him to come and have a look for himself. He was not happy. Two days later, they were pulling it down and a different foreman was on site. He was Irish and said to me, 'Jesus, it will be right next time,' and it was. There was not much kit in the shed, so I asked Bob to make some hurdles at the sawmill. They made a heap; we split the shed with sheep cratches and fetched a small shed for me to have for making tea, an armchair to sit in, and we were ready. It was a bit Heath Robinson but better than outside, especially for the Half-breds.

On 9 March the *Farmers Weekly* hedging and drainage competitions were to be held at Frome in Somerset. I told Bob I wanted the day off and he said that the digger driver, John Moss, might want to come along. I saw him later in the day and he wanted to join me, so I said we would go on Friday and stay the night somewhere handy. On the morning of 8 March there was ten inches of snow, so I rang John and let him know we wouldn't go until the evening. I dug my car out and we went about 5.30pm, and as we travelled down the snow disappeared, but it was hellish cold. We reached just outside Frome by 10pm and stopped at a pub on the side of the road. I went in and asked the old lady had she got a bed. She said she had one double if we didn't mind sharing. I said that was fine and she asked if we wanted some food, so we were lucky we didn't have to sleep in the car as it was so cold. We went to bed about 10.30pm and had a poor night's sleep because a railway line passed near the pub and a train went past every hour. We were up by 7am and had breakfast before moving on to the match. I was given a poor length of hedge with a six-inch drain coming out halfway along it, so I had to ditch it in to get it to look alright. I only had enough wood to make a hedge, so I had to make it last and just managed to get finished, with no wood left at all. It looked OK compared to the seven others around me and I hoped for the best. I won first prize and I was so pleased. Of the seven first prizes in the match, six went back to Wales and only one to a Midland-style hedger. We drove home that night to Wales and arrived back at 10pm, so it was a long day.

It was now time to put the ewes in the shed and get ready for lambing. We put the wooden troughs and the hurdles in; we made some pens around the outside of the shed and fitted an electric cable into my shed for my electric fire, the light and kettle. The sheep looked well in the shed. When we started lambing, I worked thirty-six hours on and had a night's sleep. John Griffiths covered that night and I was back in the shed at 6.30am. When we got going, they lambed pretty quickly, which kept me busy. The hardest job was getting the sheep out, but we managed after a couple of weeks.

A few weeks later I was in the shed one night, about 11.30pm, having a cup of tea when the door of the shed was ripped open and Lord Newborough stood there as red as a beetroot. He had been to a dinner in Ruthin and came past to try and catch me sleeping. He said there were two ewes lambed out in the shed. I told him that I knew and I was waiting for them to have a second lamb – then I would pen them up. He could see I was not very pleased and went out the door.

The night before 1 April it started snowing about 6pm and blew a gale, so there was a mess the next morning. John

Back to Wales

Griffiths managed to get up to the shed in his car and I asked him to take over when I went for breakfast. When I came back to the shed, we went to check the lambs that were out. There was a lot of gorse around the hedges and the snow had drifted over it and killed a lot of lambs. We picked up 300 over the next few days and there was nothing we could do about it. The rest of the lambing went OK, and when we marked the lambs I was pleased to find we still had 150 per cent with Speckles and 160 per cent with the Half-breds – without that snow we would have had a terrific lambing.

During our time at the Rhug, Lord Newborough's son had his eighteenth birthday. It was a big job in the shed with loads of guests. We had a good night and he now became Lord Newborough. On the Sunday night, a few of his young friends went over to Belan Fort and fired the cannons – there were about eight of them – across the Menai Strait. One cannon went through the sail of a ship, so it was nearly a disaster! Boys will be boys!

Bob McClain came up a few days later and said Lord Newborough was not happy with the lambing percentage. I told Bob that he'd got a bloody short memory! I was not happy.

Later that week a salesman from Young's Sheep Dip came by and told me I wouldn't see him again in that job as he was leaving to start his own company. I asked who was doing his job then, and he said it hadn't been advertised yet. I asked him if I could do it and he said 'No problem', and did I want him to give my name to the boss? I said yes. The boss rang me up three or four days later and asked me to meet him in the car park in Corwen – that's when I met the biggest character I've ever met! He was six-foot four-inches tall and twenty-four-and-a-half stone. He asked me what I had been doing, I told him and after a while said 'Look here, boy, if you can work for Lord Newborough, you can work for Young's', and asked when I would want to start. I told him I had to give one month's notice and he told me to go to the office in the morning and put my notice in. I went down the office and Bob Mack was there. When I told him to put a month's notice on the book, he said 'You don't mean it, do you?' I said 'Yes, I do. I don't work for anyone that doesn't appreciate what you do, so that's that.' He didn't like it, but I meant it.

CHAPTER 13

Lucky for some...

THE NEXT JOB was to find a house, so we bought a new house on a new estate in Ruthin – it was the only one left, number thirteen. I said to Marg, it won't be number thirteen for long, so we put a name on it. Once again, we loaded our stuff up and moved out.

I had to go to Caersws to have a week's training with my boss, Jack Parsons. He was a good salesman and hated leaving a farm without selling something. At night he took me to the local pub. That was training as well – he could drink like a fish. He told me his lifestory on those nights. As a youngster, he was a professional footballer for Yeovil Town, then he joined the army as a cook, and that's when he put on weight. He also took up all-in wrestling and fought some of the best in the sport.

After the war, Jack met a friend of his from London who told him there was a real shortage of cups and saucers and, if he could find some, to take them to Petticoat Lane where he would make some cash. He rang a friend in the Potteries and asked if he knew of any seconds for sale. This chap gave him the names of two different places, so he got on the phone and they told him to come up and have a look. The first place he visited had Nissen huts full of seconds' cups, plates and saucers and he bought them cheap. He got his friend to find some women to wash them, bought a big van, loaded it up and went to Petticoat Lane. The manager there was a bit funny with him to start with, but he got a pitch and sold the vanload that day. He booked the pitch for the following week and did the same again. The rules were that you stayed within a chalked line, but he wanted more room. So he paid a policemen to come to his stand where he had his stuff outside the line. People kicked up a row and, of course, crowds came from everywhere to see what was going on! When he had a large crowd, the policeman slipped away and Jack sold his load in no time. The same thing happened every week and he made a lot of money doing that. Afterwards, he took the job with Young's and worked his way up to become the head man in Wales.

My area was Denbigh, Flint, and Anglesey. I started with a list of customers in each area and soon got to know a lot of people. I had some very good customers who bought all year around and some I only saw once a year. They paid their bill for last year and ordered what they wanted for the next year – that's how it was on many farms. My best customers were people like Captain Archdale of Nannerch Farm, Mold, who would ring me with an order for different things in the drench line and tell me to be there by 8.30am for breakfast. When I arrived, he would have it ready. He had everything off me. I once looked after the farm from a Wednesday till a Sunday as he wanted to go salmon fishing in Ireland. He had a very large flock of good sheep and also a large herd of cows. My orders were: 'If anything died, don't ring him, ring Cluttons.' Luckily nothing did die.

Going to Anglesey one or two days a week was hard work. I had to get away early as it was a hard drive through Betws-y-Coed. Gwyn, another salesman, lived there and he was quite a character too. I used to call on my way through for a cup of coffee. I'd go to his door and shout, but he would still be in bed and he'd shout down to put the kettle on. He had started selling with Young's when he was sixteen. His father had been selling for them and he had died a young man, so Young's allowed Gwyn to take his place. His mother had to drive him round, but he managed alright. He had a big run and a lot of good customers. Gwyn wanted to learn to shear, so I taught him on a farm near Llangollen one Saturday. It nearly killed him because he was not used to hard work, but he soon learnt it and he enjoyed it.

Sometimes he would come with me to Anglesey for the day. He was a good man with mountain ponies and there were some good pony studs on Anglesey, so we would call round. We spent some good days looking at ponies and selling them some dip at the same time – most of the studs also had flocks of sheep.

Anglesey was a hard place for sales. The real Anglesey farmers were very tight with their money, but I had some good customers who moved on to the island from Montgomeryshire and off Snowdon. One day I visited a small farm to collect a bill and had a job to find the place. It was down an overgrown lane, but I got there and knocked on the door of a small house. An old gentleman opened the door and asked me in. His wife was sat in the armchair by the fire and the carpet in front of the fire was covered in cats – the room had a horrible stink. The window was open about six inches and cats came in and out across the table. He offered me a cup of tea, but I said 'No, thank you', collected the bill and went out for some fresh air.

I had a job to find my way around Anglesey as out on the country roads there were no signposts. You would often come to a crossroads with no signs at all. A good area for me was around Llangollen, and up the valley there was a farm called Tregeiriog where a Captain Greenwell farmed. Up that valley I had every farm on my book except that one. I called regularly but all I got was, 'No, I'm OK at the moment.' One day I was following him down the road with a bunch of heifers and I could see that they needed a drench, so I wound down the window and told him he'd got a nice bunch of heifers there but they would need a drench soon. He said yes, so I suggested to him that he used Nilverm injection as it had just come out, and that if he wanted them done, I would inject them on the Saturday. He said OK, and to bring enough for about 100. I went on the Saturday and soon injected the cattle. He then asked what I was doing the next Saturday. I said not a lot and asked why. He'd got another 100 to do on his farm near Wrexham, so I went and did them and he was very pleased. After that I had all of his business for everything. The following June he asked if I knew anyone that could shear his sheep, as his contractor had retired, so I got Medwyn Ellis to do the job. I was shearing with Medwyn evenings and weekends.

Another good area for me was around Corwen because the farmers knew me. One place I went to was a farm called Bryntangor, where there was a great lady farming. When I first called I told her who I was and asked her if she needed any dip or drench. She said her lambs were ready for a drench but she didn't know how to get them done as her workman was getting old and her daughter was only sixteen. She had lost her husband at the age of forty-six. He had been shearing the day before, but the next morning he was dead in bed when she woke up. She had come from Northampton as a Land Girl and met him, married and moved to Bryntangor Farm. I told her that if she wanted them done I would do them at the weekend. Again, I had all the business on the farm and I also did other things to help out. She was very grateful – there was a lot of work on 330 acres, so there was always something to do. I was getting well-known around the area and had a good business going.

I also picked up the shearing contract there as well for Medwyn Ellis – I took my holiday shearing for him. We set a bit of a record on one farm in Cerrigydrudion where we sheared seventy sheep each in eighty minutes, starting about 9pm. The farmer wanted them done as he had broken his leg. We had to fill the pen as well. They had been walked through the river and sheared lovely. Medwyn was a good shearer – he had sheared for Wales – and I learnt a lot from him.

In one of the valleys up from Llangollen there were seven farmers who turned sheep out to the mountain and had trouble with ticks. So I told them I had a new dip out that would solve the problem. I told them that I would see if I could get the first dipping done for free. I rang Professor Treeby, the man who invented it, to see if we could do it for free. He agreed, so I contacted the boss and ordered enough to dip all of the sheep. Professor Treeby asked me to be there to see it was done properly, so I went and helped them.

It was the beginning of May and these sheep had to be dipped twice, four weeks apart, because a lamb changes all of its skin in that time. By then all the dip you put on the first time was gone, and the ticks would be active again. We dipped all of the sheep and turned them out to the hill. I told them they had to be dipped at the end of thirty days, but when the time came it poured with rain and we were unable to do it until thirty-two days. I went to help gather them in and as we came down I noticed a couple of dead lambs – in all, there were eleven dead. I explained that if we could have had it done on time, that would not have happened, so we dipped again and turned them back. I called again in six weeks, and they had only lost one more, so they were pleased. The following year they all ordered dip.

One Friday, my boss Jack Parsons rang up to see what sort of week I'd had. I said it was OK but he told me he'd had a bad week – he'd had the biggest order for dip in Wales cancelled. It was from Evans, Coed Owen, Cwm Taf, so I told him I was going down to the in-laws in Clyro for the weekend and offered to go and see him. He said if I could get the order back, he would fly me in a helicopter. I told him I would go on the Saturday morning.

I arrived at Coed Owen at about 10.30am. Mrs Evans knew me from shearing there with Garney Morris, but her husband wasn't there as he was in Penderyn at a sheep sale. I said I'd go and see him there. I walked into the sale and soon saw Mr Evans and he was pleased to meet me again. I told him I had been sent to see him by the boss of Young's in Wales. He told me they had originally sent some young salesman who had given him a load of cheek, so he had ordered him out of the yard. Mr Evans ran the biggest flock of sheep in Wales, about 7,000 grazing up on Pen y Fan and good white-faced sheep, so the order was large. Most farmers ordered in five-gallon drums, his order was in forty-five-gallon drums. He said that since I'd gone down there, he would order it with me – and he ordered winter and summer dip for the next year as well. We went home on the Sunday night and I rang Jack. He was pleased and said I could put it all in for a new customer bonus – which paid for our weekend.

Every month, Young's sent out a *Young's Magazine* to customers. Jack rang me and said 'We're famous, boy.' We're on the front page! He was top salesman for selling drench and sundries; I was top salesman for dip.

One Friday night a little later, Jack's wife rang to say he was very ill in Machynlleth hospital. When I went to see him he could hardly talk and said he thought his days were numbered. It was hard seeing him so ill; we had become good friends in a very short time. On the following Tuesday, his wife rang to say that he had died, his heart had given out. I expect carrying a lot of weight was a factor, but he had lived a good life, he was about sixty-four. The funeral was a huge affair, he was very well regarded.

About six weeks later, head office in Glasgow rang to see if I would go down to Jack's area in Montgomeryshire and visit the best customers as there was a lot of money out and orders to collect. They had sent me a list of customers, so I picked seven or eight in one area, and the following Monday I got up early and went down. I called at the first farm at about 9.30am and a lady answered. When I told her I was from

Young's she burst into tears and said what a wonderful man Jack was. The farmer soon came and said the same thing. I was asked in for coffee and had to have food. I collected their cheque and was given a good order for the year. I did six more calls that day but I couldn't rush the job because they all wanted to talk about Jack. I had to have food at each farm and collected cheques. They all ordered for the next year – it was a very good day for me and Young's. I did the same thing the following day and ended up doing two days a week down there until I cleared the list.

CHAPTER 14

The farming life once more

JEFF DAVIES AND I had kept in contact over the years, and one night he rang up and asked me if I wanted to have another go at farming. There were two farms for sale together and he wanted the one, but he couldn't afford both and he asked me to go down and have a look. We went down the following weekend and walked it and decided to have a go. Jeff and I agreed a price we would go to and that I would have to go and buy it at the auction in Brecon. I went and was sat in the front, and ended up buying the two places for £150 over the price we had agreed with the bank manager in Hay. I rang him, and he was very pleased, so we now owned Pant Gwyn, Merthyr Cynog.

We had a flock of 180 ewes, sixty hogs and four rams and also had some hay in the barn. We had to put our house on the market in Ruthin but we didn't sell it for about six months. I gave in my notice in at Young's, and we moved down in November. The day we moved we had no water, as the tap was frozen outside the house. It was very hard for Marg coming from a new house to a house without a bathroom and running water, but we soon made it better. When we went to light the Rayburn, it was full of jackdaw nests. I had five or six barrowloads out of it and the chimney. The house had not been lived in for seven years, so the first

The farming life once more

job was to put a bathroom in and sink unit downstairs. We put a septic tank in below the house – I did that myself with the help of a local man with a digger. We also put a bathroom, a sink unit and toilet upstairs, along with an airing cupboard and immersion heater so that we had hot water, which was a big improvement.

We did a lot of wallpapering – two rooms downstairs and two upstairs to start with. The ceilings were high and took a bit of work, but it soon started to look like home and, after living in a town, the kids enjoyed having plenty of room to run around.

Jeff had advised me to go into dairying as it meant receiving a regular cheque, so that's what we did. There was a tidy cowshed, so we put in a six-point milking machine in there, though it was a bit of work plastering the walls and whitewashing it to be clean. We then built a shed on the end for the pump. It was a modern machine with five-gallon glass bottles above the cows that passed the milk through to the shed to go into churns. After it was finished the cows arrived in the second week of March – eight Ayrshire cows from Kenneth Beeston in Cheshire, and two weeks later, another eight. I was having to learn a new job but soon got used to it. One of the second lot of eight cows hadn't calved, and when she did I tried to milk her. It was impossible to touch her, she would kick me from anywhere. Jeff came over and we had another go, but it was hopeless. He told me to ring Beeston and tell him she was a waste of time, which I did, and he told me to sell her and her calf and he would refund some money.

Soon, lambing started and we had good weather, so we had a good amount. The Ayrshire cows stayed out. I hadn't got sheds for any more cows, so I looked into building a new shed to make it easier for cleaning out and feeding. I managed to get a grant to do that and lay a concrete yard, so when autumn came I set about getting the site ready. I hired an old Drott dozer and levelled the site. It was a bit wet and we had to put a few loads of stone in the worst places. The shed was a cow kennel type that would hold thirty-eight cows. This done, we had to concrete the alleyway and make the cubicles and fill them with sand and chippings. To make the yard, I had to concrete block three sides. This was a lot of work, but Jeff came and gave me a hand.

During the summer I got to know my neighbours and went to help Doug Wilson next-door with the harvest. He made a lot of bales and sent his workman to bale for me, so it worked out well. He had a lot of sheep in the range at the back of us, so when we turned the sheep out he was the one who helped to gather them in. I also went there to help with the shearing – he had a large flock of Radnor sheep.

I also helped Jeff with his shearing and went with him contracting on several farms. Once we finished the work on the yard and the shed, I decided to go to Carmarthen to buy some Friesian heifers. This was a bit of a job as I hadn't got too much money after doing all the work on the farm. Anyway, down I went and I was talking to Bob Jones, the auctioneer, before the sale and he said he would support me with a loan that could be paid out of the milk cheque every month. I agreed to buy eight that day and another eight in a fortnight, and he was happy with that. The sale began and I started buying. Being new to the market, the big buyers wanted to know how many I wanted. One buyer was Dodd from Cheshire. He was such a big man, about twenty-five stone, they gave him a special seat. He sent his driver to see me. I said I wanted eight and didn't want the posh ones and, fair play, he didn't run me too much. Another chap called Speak was from Cornwall, but he didn't want the sort I wanted. I required hardy heifers, Pant Gwyn was above 1,000 feet. I bought eight and got them home. I had talked to the people I had bought them off, and according to them they had all been milked. Boy, did I have a job milking them

The farming life once more

for the first week! My neck and shoulders were as sore as if I had been in a rugby scrum from holding square to put the machines on. They settled down, but the next eight were just the same – they'll tell you anything to sell down in that market! The best heifer that day had a lovely udder and the lady owner said she was the best heifer they had sold, and she was almost crying! When I milked her out, her one back teat went right up half the size of the other one.

I was now milking thirty-one cows and turning out a lot of milk. In the house there were two dairies side-by-side, one had salting stones along one side for salting a pig and in between was a wall. We wanted to make this into one room, so the wall had to come out. I told Jeff about it and one night after milking he brought his tipping trailer. We had a job! It was a stone wall, about eighteen inches thick, and we had to make a hole through it to start, which took some work. We had four trailer-loads of stone out of there. When it was done, it made a good room and we decided to make it into a kitchen. The first job was to put plasterboard on the ceiling where the wall had been and plaster to match the other ceilings. We gave it a day or two to dry, then one night we papered the ceiling and finished about 10.30pm. When we got up the next morning, I looked in to see what it looked like. The whole lot had come off – apart from the new plaster down the middle! Margaret was mad, after all that work. We had a think about it and decided to put a sealer on it first and did it again. This time, it stayed up!

We then put a good kitchen in with worktops, sink unit and cooker and painted the walls and put lino on the floor – it looked real smart.

I needed more grass the next spring for thirty-one cows. When a man from Dunes Fertilizer came by, I asked him what was best to do and he advised me to put some of their fertiliser on the best and lower fields, then put some forty-six per cent nitrogen on at the end of February and I would

95

have grass. I put their fertiliser down on Boxing Day as it was fine, and then put forty-six per cent down in February. He was right, we had plenty of grass and I turned the cows out early.

I was using AI on the cows and had about five that wouldn't go in calf, so my brother-in-law offered me a bull calf to run with them for a month. He came on 13 May, and when we took the bull calf down to the cows he went straight through my electric fence. My brother-in-law had a chap from Woolhope near Hereford with him who was also in dairying. When we went into the grass to get the bull calf back, we were in well above our knees. He said I'd got more grass there than he had at home. I had to mow some of it at the beginning of June.

Jeff had started improving his farm, which had a lot of rushes, so he drained it and improved it. He was now turning sheep out on the range and I gave him a hand to put a roof over his silage pit in the winter. In the autumn he laid a concrete road from outside his cattle shed.

One Saturday I went to Talgarth Hedging Match, and about halfway down my length I cut my leg with my axe. As I was going to cut a cropper off and swing my seven-pound axe back, it caught on a briar which turned the axe in my hands and into the side of my leg, cutting it the full width of the axe. It went in about quarter of an inch at the deepest point. One of the judges was stood by me when I did it. I looked at it and said 'It's not too bad', but he told me go to hospital. I went to Brecon hospital on my way home and a shirty nurse said to me, 'What's the matter with you?' When I showed her, she took me with her and they stitched me up and gave me some pills, and home I went.

We were doing Jeff's concreting on the following Tuesday. We started after milking, at about 9.30am, and went on till evening milking. Then we started again after milking and finished at 10.30pm. I was mixing and Jeff was laying. By the

The farming life once more

time I had finished my leg was very painful but it got better in a day or two.

One job that needed doing to the house was to put two new windows in each side of the front upstairs. They were sash windows, but the lintels had gone. I had an old oak tree in a small wood and took it down to the sawmills in Battle and had new lintels made. The metal windows I had bought were eight inches too wide, but the postman was good at welding so he took eight inches out of each side, making them a matching pair. I took the old windows out and fitted the new lintels. The windows were the hardest job – rebuilding a small wall above the lintels and glazing them – but I managed to get it done. It certainly improved the look of the front of the house.

Above Pant Gwyn was a Scots pine windbreak which was overgrown. There were some fair trees there, so I decided to fell them and make a bit of money. I bought a chainsaw and started felling and pulling the timber down to where it could be loaded and I sold it to Sennybridge sawmills. It took a bit of time but it looked a lot better.

We had a small Caravan Club site on a piece of land near the house – which brought in a bit of cash in the summer – but they were not all good campers. We had a water tap on a tree for their use, and one day when I went past a young woman was washing dirty nappies out under the tap. I told her off, I can tell you!

I was using an A30 van to take the milk out. Now that I had more milk to carry, the springs went. I told John Williams, a farmer just down by Upper Chapel who did a bit of mechanics, and he said he knew of a Morris 1000 van with a knackered engine going for £15. He said she was in good nick, so he bought it and swapped the engine out of my van, so I was away again. He was a good chap if you were in trouble.

During the summer I bulldozed the middle out of a small

wood, leaving twenty yards for shade at the top and thirty yards of steep bank at the bottom. Then we ploughed and reseeded it and it came in very handy for shade for young lambs.

CHAPTER 15

On the move again

IT WAS 1974 and I was now getting itchy feet again as there was no chance to get more land around Pant Gwyn. In August I saw an advert for a farm to let in Beulah, with 178 acres of land. Marg was not happy because she was working in Brecon as a wages clerk and would have to change her job. I said we were not sure to get it but I wanted to expand, as sixty-four acres was not enough to make a living.

The Labour leader Harold Wilson had been Prime Minister from the time we left the Pant, Rhulen, to when we moved to Pant Gwyn. Nothing had changed – land, sheep and cattle were the same price. Ted Heath was now taking us into the Common Market, so we hoped things would improve. We got down to the last six for the farm at Aberannell, and I went for an interview in Builth with Charles Woosnam and Mervyn Bourdillon. I went up to the office and told them who I was, but Charles Woosnam started firing questions and never let me answer properly. As I came down the stairs a local couple were going up. I told them 'That's a waste of time', and went home. There were seventy-six people trying for the farm.

After I'd told Marg about it, I went round to Doug Wilson's for two sheep he'd brought off the hill for me. When I arrived Doug was at the house door telling me to come for a cup of tea, and asked how I had got on. I told him it was hopeless. He said that local people hoped that I didn't get it as they wanted me to stay where I was, which of good of him. Halfway through my cup of tea the phone went. It was Marg telling me to come home as Woosnam had been on the phone, so home I went and rang him. He told me to be in the office on Monday to sign for the farm. I couldn't believe it!

We now had to sell Pant Gwyn, which we did in two lots – the house and eleven acres plus some buildings, and then the remainder of the land, fifty-three acres. We made more on the house and eleven acres than we paid for the whole farm four years earlier, so the land was a bonus, but we had to buy a good flock of Beulahs for Aberannell and we needed the money. The barn was full of hay, so we bought four hundred Beulah ewes and 120 hogs, some rams and twenty Welsh ewes as well.

The house at Aberannell was alright to move into, but the tenant wanted some time to get his bungalow finished, so I took the farm over at the end of September. I had bought a petrol Land Rover and a twelve-foot Ifor Williams box, and I started moving some dry cows over. One Saturday I had six Ayrshire cows on board, and on my way over from Upper Chapel I came down above Corrin Farm where there was a bad left-hand corner. Sixty yards before it, I put my foot on the brake to slow down and it went straight to the floor. The kids were along with me, so I told them to hold tight as this wasn't good, the brakes had failed! I went into the corner and managed to get around, just brushing the hedge. The Land Rover was gaining speed down to the next corner, but I got around that one as well. It slowed down a bit over the crossroads and we went into the next corner. We were now getting close to crossing the old river bridge where only one car could pass, so I hoped nobody would come. The Land Rover was singing by the time we reached the bridge. I aimed for the middle and made it through. The Evans luck had held out!

I got to the main road and drove very slowly to Aberannell. After I let the cows into the field by the road, I left the box on the yard and drove carefully to Prynne's garage, pulled in

On the move again

and asked Tony Prynne if he could fix it. He said he could and gave me an old car to go home in. When I told Marg she couldn't believe it, but I don't think the kids realised how much danger we were in.

That year was bad and we never made any hay until August. It was old and strong by then but there was plenty of it, so I had to move it down to Aberannell. My neighbours all came with big trailers and helped. Gomer Probert next-door had the biggest one, so he took the heaviest load, but the rest loaded up well and we moved it all in a day. I was very grateful for their help.

We didn't finally move down till the Christmas and settled in well. As soon as we moved in the cows started calving, so I was buying, and putting two calves on the milking cows. In the December sale, the cattle were a bad trade. The poor summer hadn't helped and things were tough, but it suited me because I wanted to buy some bulling heifers. I bought seven for about £50 each, also two cows and calves – a black heifer and calf cost £105 and the other was a Blue Grey cow with a six-month-old heifer calf costing £76. She was a bit of an age but very cheap. The following Wednesday in Builth, I bought six black Hereford heifers for £67 that were ready to bull.

There were hundreds of Hereford calves just weaned being sold for £7 and £9. In December I went to Llandovery and bought fourteen Charolais heifers for £16 each, which was very cheap. I kept them until June and made a good profit because the trade had improved by then. A friend of mine took a bunch of good heifers to Hay-on-Wye in September and was offered £57 each. He didn't sell and took them back in December when he was offered £27 each. He didn't sell again and had to keep them until the following June.

In February and March my cows were calving well, and by April I had reared seventy-six calves which would need a lot of grass. The next farm, Celsau, put out grass to let, as the tenant had retired. I rang Charles Woosnam and managed to rent the seventy acres, so it made all the difference. I put all the cattle over there for the summer and we were able to make plenty of hay for the following winter.

The children were able to go to Beulah school. Margaret got a job as a carer in the area and she was enjoying it.

I changed my Land Rover for a newer one and bought a Zetor 3545 four-wheel drive tractor as some of Aberannell's land was steep and needed improving. By then there was a new grant scheme, the Farm and Horticultural Development Scheme (FHDS). I had a man from the Ministry out and put in for a lot of hedging and fencing, plus some draining and reseeding. That first year I ploughed the piece facing Beulah church. It was very steep but with four-wheel drive I had no problems. Some of the field had been fir trees and had never been ploughed before. I was ploughing up stumps across the steepest bit, but it came out well.

The Speckle ewes lambed well and with all the extra room the lambs thrived too.

In the autumn I was asked to go to an NFU meeting in Builth and enjoyed it very much. My training in public speaking at Young Farmers was put to good use and I started speaking up for farming. We had some very good debates. I was also asked to join Builth's Ploughing and Hedging Society Committee, which put on a ploughing and hedging match in April every year. I knew most of the people, so that was good. Also, Gwyn Edwards, the tenant before me, was a member of the Cefnmaes Trust in Abergwesyn and asked me to take over from him, so I did.

When we started marking lambs at the beginning of May, I asked Gwyn if he would show me how to do the earmark on the lambs because it was a difficult one to do. He came and stayed to mark all of the lambs. When we were doing it I told Gwyn that, as part of tenancy, I was supposed to resettle the flock on the mountain as he had stopped using the hill for

On the move again

twenty years. He said that would be hopeless unless I got the shepherd on the hill, John Lewis, Pencaer, on my side, as he was paid by the estate to do the job anyway.

Gwyn and I were talking one day, and I said I had been at the Dai Felix, Trallwm sale and bought a heifer which had a halter on her. Gwyn said that wasn't Dai's heifer, it belonged to John Lewis who had led it down to the sale on the halter. He said if I told him that when I saw him, I would have a friend.

One day I was going down to meet the kids from whist drives and Young Farmers, which I did every week, and waited at the Trout Inn and had a pint at the same time. John Lewis was always there – he used to walk down the three miles from his cottage every night. I got my pint and sat down by him and told him I had bought a heifer off him at Trallwm sale. He asked which heifer, and I told him it was the one he had led down on a halter. I explained that my father and Dai were good friends and he had bought a cow and calf. John asked, 'Duw, who's your father?' I told him he was Harry Evans, Llwyn Tudor. John asked whether he sheared with Ernie Mills up the valley and I confirmed that. 'Duw, I knew your father,' he said. Before that he would never speak much, but suddenly he was up to the bar asking for a pint for Evans, and when the kids came in from YFC, he bought them some chocolate. From then on we were friends.

For the first week or two I didn't say anything to John about putting sheep out, and when I did he said 'Duw, there's been no sheep out from Aberannell for years.' He said he would see to them as long as I stayed off the hill. We had a right for 500 sheep, so there was plenty of room. In the autumn he asked when I wanted the sheep down. I said 'When you gather your own and don't gather special for me.' He promised to do it the following Thursday, and sure enough the sheep arrived on the yard at 6.30pm. He had brought them down on the way to the pub!

103

In the autumn I decided to take two lorry-loads of weaned calves to Hereford market, and when we had sorted them out I was stood by my first pen of nineteen Black steers. They looked well. The auctioneer came by to have a look and said there would be no problem selling them. As we were talking, a large, grey-haired man came over, had a look and went on. The auctioneer said 'That's your buyer.' He was right. He was a Mr Matthews from Bromyard and he paid £100 each. They had done well – they cost £30 as calves in the spring and I had a £12 sub on each of them. The rest of them also sold well, so my hard work in the spring had paid off.

I went down to Monmouth to buy a Shorthorn bull to try to breed some Blue Grey heifers, but it did not work too well. We had too many bull calves and only a small number of heifers, so we sold him on.

In my tenancy I had to have Beulah Show on the fields down near the village on the second Saturday in September every year, and I was asked to join the committee. We had two meetings a year. When I joined it was pretty old-fashioned and just a show, now we have trotting as well. Some of the old chaps on the committee would have been to Builth market beforehand, as the meetings were held on a Monday night. They would have a few pints in the Barley Mow pub before coming home and were a bit hungover. They were bound to be falling out about something or other before the meeting was over. I've been on the committee now for forty-three years and it is still the same.

The winter going into 1976 was dry and easy. When spring came it was the same. I put fertiliser down at the end of April and it was still on the ground when we cut the hay. The summer was the driest and hottest summer we had here. The hay crops were very light and very dry, you could bale it in twenty-four hours. It was not even half a crop.

After weaning, I had the ewes running above the house and the gate open into the forty-five-acre wood. The ewes spent the day lying in the wood and would only come out to graze at night. There was only one place for the sheep to drink in the valley, under the wood where there was some water under a rock about as big as a bucket, but they managed. The Annell brook was dry for two months.

I had the cows over on the top side of Celsau again for a while, but the water dried up. So I put them on the meadows below the road that had a river on two sides, as water was very important. I had hired a Charolais bull off a chap near Bromyard and he did his job alright. He was a bit of a boy. When you went to see the cows, he would round them up like a stallion does with mares and would not let you into the cows to check them. I used to have to drive the Land Rover in among them. When we had to move them he was a bit nasty, so when Mr Godwin, the man I hired the bull from, came for him I told him to watch him.

At the end of August we went on holiday to Borth in a caravan. We had a good week and came back on 3 September and ran into rain in Newbridge. It was magic! But once it started raining, it kept raining. We had a wet Beulah Show, but nobody complained. The grass started to come and kept growing until January because the ground was still so warm.

The ewes were pretty poor at the end of August but had put on some meat by Christmas. During the summer I had been talking to Mr Beynon, Ty'n y Cwm, and had mentioned that one day we were going to be short of hay. He suggested that we should go up to the top of Ty'n y Cwm and cut some molinia grass, as it was good cow fodder. I went up to look and there were twenty acres that could be cut because it was so dry. Normally, it wasn't worth going there because it was too wet. I went up and cut a good piece. I had a rotary mower and had to go in bottom gear because it was tough to cut. I left it three days and went and rowed it up and baled it, which was very hard, but I managed it. The track up there from our end was a bit hairy and only just wide enough for my trailer, and in one place I had to drive over solid, slippery rock. I loaded up about 100 bales and roped them well.

Gwyn Edwards had come with me to load, and on the way down with the first one he asked to be let off before we drove over the rock. There was a hell of a drop below and if things went wrong, there was no hope. The four-wheel drive on my 3545 Zetor helped and we drove down four times without a problem, but there was a shower of rain before the last load and Gwyn told me to be careful. I stood topside as I came down that bit, ready to jump at any minute if I needed to, but I just made it! Gwyn told me that at one point there had only been half the trailer tyre on the rock.

Mr Beynon wanted two loads for himself, and the road up from his end was bad, so we used his trailer which was smaller than mine. We got up there and loaded about 50 bales and roped them well. Everything was fine until we were three-quarters of the way down where we had to go across one track and turn down another. Halfway across the second track, his trailer got stuck on a stone and tipped over, breaking the wooden towbar on the trailer. We put a new drawbar on the trailer and reloaded the next day and got the second load down OK. That's when I first saw Ty'n y Cwm.

I had a bunch of cows out behind the building and I fed the molinia to them, but after a while I could see they were having trouble digesting it. I decided to give them some beet nuts to help them. This worked, and they were alright. The ones in the building had been out until Christmas and I fed them half hay and half molinia. When they started calving by the Charolais bull, the calves were coming out all colours of the rainbow – from pure white to black and brown, but they were good calves. Considering the amount of fodder we had, we got through the winter fairly cheaply.

When we got to spring we had two to three inches of snow every morning for the first week of April. It would melt in

On the move again

the daytime then snow again in the night. We were lambing out at the time, so it was hard work. I had done a fair bit of hedging in the winter and Gwyn had come to help. He was a good hedger but hadn't done much of it at Aberannell for years. He was a great help, though it all had to be fenced later. The hot summer of 1976 had burned the leys out of some of the fields, so I ploughed some up and reseeded them under the FHDS at fifty per cent and that was a good help. I also expanded the Speckle ewe flock and started selling yearling ewes. I had bought two pens of registered Speckle ewes, so they allowed me to register my flock from then on. We sold 100 yearlings to help pay the rent.

We rented Celsau again and put the cows back over there and hired another Charolais bull off Ray Goodwin. When he arrived we turned him to the cows. The next morning, I watched him. There were three cows bulling and he was having trouble serving them, but I thought he was a young bull and would soon get the knack. Three weeks later I could see he had a broken rod (this is his penis or willy). I rang Ray and informed him the bull was no good. He couldn't believe it as he'd never had a cow before. He agreed to find another bull because the cows were running out of time. He rang me the next day to say the only bull he had got was a full French bull and I would have to look after him as he was worth a lot of money. When he came, he was a monster of a bull, well over a ton in weight, and was working well. He went lame with foul after a fortnight, so I got stuff from the vet and went down to inject him. I wasn't sure if he would let me do it in the field. I filled up the syringe, went up behind him, stroked him and banged in the needle and he never moved, so I filled it again and yet again, he never moved. He was better in two days. I did wonder how my cows would calve after him, but they were fine.

The dipping facilities at Aberannell were very poor, so we knocked them down and put up a new set of pens and

107

a round pen to push the sheep nearer to the tub. We left the old tub, as it was brick-built and in good nick and put in a double-drainer, which also speeded things up. We were having to dip twice a year with **OP** dip to stop scab, which had become a problem. I was able to dip 700 to 800 sheep a day on my own with the new system, which we'd had an **FHDS** grant to install. I also applied to put new stone tracks around the farm, a silage pit up in the oak wood, and also a new shed to join the cowsheds to the Dutch barn. The tracks were done first in August.

CHAPTER 16

Ty'n y Cwm

I HAD HEARD that Mr Beynon, Ty'n y Cwm, was retiring so I rang Charles Woosnam to ask about it. All he said was, 'Your interest is noted', and goodbye! During the summer some land came up for sale at Beulah. It was twenty-seven acres of rough land, so I made enquiries and bought it for £6,500. I intended to improve it and the first job was to bulldoze some trees and a hedgerow out to make it easier to plough and drain some wet eyes (small bogs) out of it. I had Gerald Powell look at it and he said he could do the job. We laid four-inch pipes on an angle to catch these wet pieces, dug the eyes out and filled them with river stone from a river that went through the corner adjoining the village. We did a good job and drained a boggy piece near the village. I left it to drain over the winter.

Mid-September, Charles Woosnam rang and asked me to be in the office to see him about Ty'n y Cwm – he offered it to me on a yearly rent. I would have to buy the sheep flock, so I agreed a rent and went ahead and bought 400 ewes and 280 lambs, including ewe lambs and some rams – a hardy Speckle flock. Ty'n y Cwm had a hill for 900 sheep, and it was a settled flock which suited me grand.

The handling pens at Ty'n y Cwm were hopeless; how they had managed, I don't know. At the time, the holding was 330 acres with 200 acres growing bracken and molinia. It was going to be a challenge, but I was looking forward to it.

Where The Hell's The Time Gone?

At Aberannell we put the new shed between the cowshed and Dutch barn. I wanted to take out the cattle stalls and make it into an open shed so that the two, side by side, would hold the cows for the winter.

One day I was taking some sheep up along the road to Ty'n y Cwm and a car came up behind me. The driver wound the window down and said hello. I asked him if he was the man with an advert in the *Brecon & Radnor Express* looking for work. He said he wasn't, but he did need a job. His name was Peter and he was from London and had been in college, but he hadn't ever worked on a farm. I told him that I had a lot of work to do that winter and needed some help. I couldn't pay too much to start with, but I would train him a bit, and he asked when he could start. He started the following Monday and, fair play, he was not afraid of work and never moaned no matter how hard it was. We started sending our hogs off to tack for the winter, so with Ty'n y Cwm we needed tack for about 400. It cost a bit but they grew better away on good land.

We also started working on the cowshed and the new shed. The bottom end of the shed had to be concrete-blocked, so we did that and blocked under the feed barrier, then tackled the cowshed. It was a lot of work, but once it was done it made an excellent shed. We concreted it all out and it was a great improvement.

The big problem with the land at the top of Ty'n y Cwm was access. You had to take a tractor up to feed the top because you had to cross a bog. So I decided to feed the ewes Rumevite blocks to save going up every day and that worked OK. The Beynons hadn't drenched or injected the ewes with anything, so I drenched and injected them for fluke and worm, and then injected them again in the spring. The Beynons had been doing alright, though they hardly had any twins – if they did, they put a cross on the calendar. Mr Beynon used to inject the lambs as they came, which meant

Ty'n y Cwm

walking up to the top twice a day for three weeks because otherwise they would die of dysentery.

Mrs Beynon ran a fish farm adjoining Ty'n y Cwm and did a good job of it. She was much younger than Mr Beynon and worked hard. They stayed in the house the first winter, and when lambing started she said she would lamb the ones I had on the meadows. I agreed. She couldn't believe how many twins there were – more than thirty pairs, as I had separated the heaviest ewes when we injected them.

In May I got Dai Kinsey to plough the land we had bought at Beulah. I went over to look the morning he started and he had gone out about 100 yards. It was going over well with plenty of soil – on the top field the soil was red and looked good. Jim Edwards, Gwyn's brother, was walking over that way to his sheep as he still had some land – he couldn't believe how good the soil was. Kinsey came and rotovated afterwards, and also put lime and slag on it while I got it ready to sow. I saw Dilwyn Williams on the road and he said he had about twelve bags of seed I could have – it was spare from reseeding coal tips down in south Wales. It was very good seed and I mixed it with ryegrass and sowed it, along with two pounds of rape and half a pound of turnips. I rolled it down hard. It grew very well and was a super improvement. The best rape was four feet high, so it lasted a good time. In August we weaned the lambs and put the lot on there and they did very well too.

I kept a few Speckle tup lambs and a couple were not doing very well, so I gave them a drench and a couple of caponising pills used on cockerels. Somebody had told me it worked – and it did. When I sold them, they were twenty-three kilos deadweight. It's against the law now.

When we got to shearing, we started about 20 June, and after a while my workman, who was helping, said he'd like to learn the job. I borrowed a machine for a few days and started showing him what to do the next day. He soon learnt

111

and worked hard to get it right. On the Saturday we had a nice bunch of Ty'n y Cwm ewes in and he sheared 130 sheep in a day – he was a good learner, I only had to show him something once.

I asked Charles Woosnam if I could plough thirty-five acres of fern on the top of Ty'n y Cwm. I told him he would have to pay for it as it was on grass keep. He agreed and said we could fence it first in the following spring.

We had a good hay harvest that year, but with a lot of stock we needed a lot of hay – 1979 into 1980 was a bad winter. It started snowing at Christmas and carried on into the new year. It had been cold through November and December so the cows were fed early and by mid-February we were short of fodder. I started buying Cotswolds hay from a Mr Jackson at £60 per ton. I'd had two loads off him, and then he rang me up one day about ten days after the second load had been delivered and asked if I would want some more. He had sent a load to Carmarthen and the chap would not take it as the top half was red clover and very stalky; the bottom half was green meadow hay. I got ready, fed my cows and got the lights on because the hay delivery would be there by 7pm. When he arrived and we took it off he was very pleased because my cows loved the clover hay, and green hay was good for sheep. I had two more loads after that and he kept the price at £60 per load, while the next-door farm was buying in hay for £90 per ton and it was no better than mine according to the workman.

It was a hard winter and it hung on. The snow went in March but then it was wet and rough through lambing. On 7 May we had a foot of snow at Ty'n y Cwm. I went up on the Bryn's, a fifty-acre piece above the house where the ewes were, and there was not a sheep in sight. I went to the far end of the patch as far as the steep wood facing the forestry and found the ewes lying by the oak trees. I cleared a path with the transport box, called them up and fed them. That was on a Wednesday. It was a nice day the next day, and a lot of the snow went but, on the Friday, it snowed another four to five inches. Finally, that marked the end of it, but that winter cost a fortune.

I saw my neighbour Bryn Powell a few days later and told him that the snow the previous week had been the last straw. He agreed that it was the straw that broke the camel's back. He had 100 ewes and had only lost two or three all winter, until the final snow came and lost thirteen. I had lost about the same out of 600. Up in Abergwesyn they buried 400 with a Hymac; the chap there didn't know how to handle it.

The last feed I took to Ty'n y Cwm was 11 June when there was still no grass but then it got warmer and things did improve. At the end of the first year, Mr Woosnam offered me a partnership with my landlord, so I investigated it and agreed on one condition – that I could put it into the FHDS and make some improvements. They agreed and I had a man from ADAS out to look and got a good scheme with a new road up to the top included, a new 100ft x 50ft shed, reseeding of all the land that grew bracken and a load of fencing.

The first job was the new road, so after I got the agreement to start, I asked Dilwyn, Cribarth, what he thought of the job and he said he would do it for the grant. The standard cost of the road was £56,000 but he would do it for £28,000, so I told him to get on with it. He sent in a big bulldozer to make the road and on the first day he reached the corner above the house. I asked the driver to make a good, wide corner and to go on up onto the fifty acres, then afterwards to go straight through the middle so I could feed off either side.

The next day I had to go to Hereford. After I came home I popped up to see how the work was progressing. He had gone straight along the side of the bank. When I asked why, he told me Charles Woosnam had come in the morning and told him to. I was hopping mad that he had gone all the way to the far end of the field; there was no way to change it. The next

113

morning I went up there to show him where to go out onto the next piece, and just as I had told him along came Charles Woosnam and Mervyn Bourdillon, my landlord. When I told them where the road was going next, Woosnam said 'No way'. He wanted it to go into the middle of a good area of land of about twelve acres and spoil another piece. I told him he wasn't spoiling another of piece of land. Mr Bourdillon asked to me walk where I wanted the road to go, so I showed him. When we came back down he told Woosnam that he agreed with me. Woosnam was mad, but I had my way and told him off about the day before – he didn't like it one little bit.

We took the road up to the top and round into the thirty-five acres I had ploughed. Dilwyn did a good job and it really made the back of the farm. We had acres we needed to get lime and slag up to and on to the reseeding we had to do.

That spring, about 15 April, a friend of mine, who had sold the Trout Inn in Beulah and bought a farm near Lampeter, rang me to ask if I wanted some grass. I asked if it was good enough for cows and he said it was, so we agreed a price and I got some local boys with two lorries and took the cows down. I went with them and there was good grass there. I was pleased, as it helped a lot because we wouldn't have grass for cows for another month.

His wife didn't like it where they moved to because she was English and complained they only spoke Welsh. In the shops they were not even willing to speak English. In the autumn, after I brought the cows home, he asked me if I would sell the piece of land I had improved by Beulah as they wanted to come back. He had kept two acres belonging to the Trout Inn and wanted to build a bungalow there. I told him I didn't want sell, but he begged me so much I agreed if it was for a good price. He told me to name my figure and I said £20,000. He agreed, as long as I bought his tractor as part of the deal. So I had a good Ford 3000 tractor and £19,000, which suited me as I had spent a lot of money at Ty'n y Cwm.

Ty'n y Cwm

In the autumn of 1980 I decided to go with Jeff Davies up to Hexham to buy some Blue Leicester rams to try on our Beulahs. I managed to buy two, one for £150 and another for £200. Jeff had seen some lambs with a farmer around Merthyr Cynog and they seemed to do well. This cross was to become the Welsh Mule, which spread like wildfire. We were among the first four or five to breed them and they have been used on Aberannell ever since. Nothing has paid better, but they are a lot of work.

About that time I also joined the Wye Valley Grassland Society, which had been going about twelve months. We had meetings every month during the winter and farm walks during the summer. I'm still a member today.

Under the FHDS scheme, I had also put in for some drainage work. I got Dai Kinsey to do one piece on the top that was so wet you could not cross it with a tractor in the summer. He did an excellent job and made it into a good field, but it is inclined to grow rushes if you don't keep them down. That summer we reseeded the thirty-five acres and twelve acres next door. It was very steep along the bottom but I had bought a new Zetor 7045 four-wheel drive tractor with double wheels that could go anywhere. It was a good tractor with plenty of power. Dai Kinsey ploughed and rotovated it and we reseeded it. From virgin land, it came well. I sowed a good ley from Nelson Love, who sold good leys for hill land and it was perfect for fattening lambs.

115

CHAPTER 17

A growing business

ABOUT THIS TIME, Jeff Davies started a shearing competition. He had built a six-stand shed for shearing his sheep and decided to have a local competition. I went over and I enjoyed going back to Merthyr Cynog and meeting the locals. My best effort there was in the local class where I managed to win the fastest time of the night – five minutes, ten seconds for five sheep. The best time that night in the open class was five minutes, forty-five seconds.

They also started a shearing match at Tyrosser, hosted by Troedrhiwdaler YFC. I had the fastest time there as well.

A couple of years later at the Pant shearing, I hadn't done very well. So I was sitting on the end of the stand, watching the open final. Merlin Roderick from Brecon, who was sponsoring the match, had been asked to do the commentary and came past to go up onto the stage. He told me he was scared and asked me to take the mike and to say something. He said he couldn't do it and would faint if he went up there. I took the mike and got the finalists so worked up it was an excellent final. I must have done a good job because they decided I was the commentator from then on. I very much enjoyed doing it and I was then asked to commentate in Llanafan as well.

After a couple of years I was asked to join the Royal Welsh Shearing Committee, along with Alan Davies, Plasnewydd, Llandovery, who was commentating down in that area. We were both asked to commentate at the Royal Welsh. He was a great chap and we got on well. We took the job, and we improved the shearing a lot over the next few years.

Back home, we had now put the silage pit up in the big wood and we were making silage off the bottom of Aberannell. This made wintering the cows much cheaper, with not having to buy litter, and the cows loved lying in the wood. It covered forty-five acres, so they had plenty of room to lie down and there was shelter from most winds. The only problem was that, when the calves were born, they needed a selenium injection because they didn't get enough sun under the trees and this affected the selenium level, so they told me. We calved the cows up there and we were pretty lucky. Sometimes we had a problem, but you could get them inside. In the best year up there, forty-four cows calved OK with only the very last one, a grey Hereford cross, having a dead calf. At the end of April, I had some help to dehorn and castrate the calves before they went to grass.

In the winter of 1982, at the beginning of February, we had the most level snow I've seen. It was about two feet six to three feet deep all over the fields and at Ty'n y Cwm. It blew, as well. I spent days making tracks for people with my four-wheel drive tractor as they had two-wheel drives and couldn't get to their stock. When it fell, it was a Friday, and on the Saturday morning about 10.30am I set off with my son Mike from Aberannell with a load of hay on the transport box to feed the sheep at Ty'n y Cwm. We went far up the Bryns above the house and got stuck. I dug around the tractor, but I couldn't get it out of there. Then I tried to carry some hay to the sheep, but with hay on your back you couldn't go far. I gave up and said to Mike we'd better head home. By now it was 2.30pm, so we walked down onto the road. Halfway home we met seven men coming up the road. Margaret had told one of them that we had been a long time, so they had got together. They had soup and tea and we sat on the ditch. I said if I could have a dream come true, a County and tow chain would come up the road. Five minutes later, that's exactly what happened! Dilwyn Williams had been called to open the road to get us out. We went back up to Ty'n y Cwm and up to the tractor. I told Dilwyn that it would only need moving a foot and it would be out. After a bit of fun, he managed to get to the back of the tractor. He hadn't brought a chain but I had plenty of baling twine and tied that to the back of my box and to the front of his tractor. He just moved a foot or two and I was away.

After thanking Dilwyn, I went down to the road and picked the boys up and went back home. Four of the men from the village said they would come and give me a hand to get the sheep down to the bottom the following day. This was no easy task. I had to make a track for the sheep to walk down and it took a few hours, but we got them down to the barn field to some shelter and gave them plenty of hay. I looked at the bunch and there were thirty to forty still missing. So the next day two men came with me and we found thirty-eight over in the next dingle and I was able to feed them.

On the Tuesday I was at the shop in Beulah and local people were saying they were worried about John Lewis, the shepherd. So five of us decided to go and see him. The next day Graham Davies, a local builder, put a pack of food in his knapsack. We took the tractor up as far as Llofftybardd turn, then had to walk the rest from there. We took turns to make a track for the rest to follow. It was a very steep road up through forestry and it was very hard work. When we got to the top we had a rest. It was just as steep going down the other side to a farm called Blaencwm. From there it was flat along, close to the river, but the snow was very deep and in places there were rushes, which made it worse. After crossing the river there was a steep pitch up to Pencaer. We got to his house and there was a ten-foot drift across the front, but the way to John's woodshed was clear. I climbed up on the drift and saw John going to the woodshed. I shouted to ask how he was. He said that he had been worse than this many times! We all went to the house and unloaded the food which Graham had carried all the way. John's face dropped when we said we hadn't got any tobacco, but we couldn't get any. We stood there talking for a few minutes and my back started to get cold, so I said to the boys that we had to go. We said goodbye and went out into the snow again – he was OK, that was the main thing.

The snow went in the next week or so, and didn't leave any mess. It didn't even seem to wet the ground. In Builth market the next week, everybody had a tale to tell. I went down to the Barley Mow for food and a pint before coming home and had a laugh with Samuel Scache, Newbridge. He was a very short man and he joked that he had to go back to the house for his wife to thaw his balls out because they were dragging in the snow all the time!

In 1982 I was asked to be chairman of Wye Valley Grassland. I enjoyed the year; we had a farm walk and when we went up around Ty'n y Cwm they were impressed with what we were doing. They could see how much work was to be done but wanted to come back when it was finished – which they did many years later. I am now one of the oldest members left.

We got stuck-in that summer and reseeded another thirty-two acres of bracken. Dai Kinsey ploughed twelve acres and I ploughed twenty acres myself. I got the twelve acres in on time, but the twenty acres was very late due to the wet weather. It was about 10 August before it dried up for a couple of days and that was late for reseeding, but I wanted it in. I rang John Saunders, who was working for Nicholls, Llandewi, at the time, to order some rye as a cover crop. He got me twenty bags and I sowed the seeds and rye and hoped for the best. When the rye came up, the field was blue for a bit then went green; it turned out well.

Getting hay that year was hard work. I got John Richards

to big-bale Ty'n y Cwm because it would not make hay down on those meadows – that was the first big-baler in the area. We had to put them in big plastic bags in those days.

After weaning the lambs we took the wether lambs onto the twelve-acre piece above the lake. It had grown well so the grazing was good. After they had finished that, I put them on the twenty acres of rye where they did very well. When I thought they were ready, I contacted FMC in Brecon and their field officer came to check and tag them. He said they could all go and they came back at £37.80, which was good.

Next, we put up a building down by the river at Ty'n y Cwm. Mogs Morgan built it and did a good job. My ewe numbers were growing, so we were able to put some Ty'n y Cwm ewes to the Blue Leicester as well. It worked and we lambed them in the shed, along with the twins. The shed was also handy for shearing because you could put them in dry and they were safe. We had pens by the shed, so it all worked well with ewes in both sheds.

I was busy at lambing time – it's a lot of work lambing ewes inside. I would get up at 5am and my first job was to check which ones had lambed and pen the twins up. I would then go to Ty'n y Cwm and do the same. Next, I had my breakfast and fed all around. I would go up to the top of Ty'n y Cwm to feed and check them and call in at Ty'n y Cwm shed to make sure everything was OK, then go home again to check there before lunch. I also had to check and feed the cows now being wintered up in the wood, as they were also calving and had to be watched. As mentioned before, the calves had to be injected with selenium. Any lambs that were strong enough went out to make room for other lambs. Mule lambs don't like rough weather, so it paid to hold them the extra day if the weather was poor. With about 900 ewes lambing in the first two weeks, it was hard work. Sometimes I wouldn't get to bed until 1am, but I had to be out by 5am. Once I overslept and didn't wake up until 7am, and seven ewes had lambed in the corner by a drinking trough. Boy did they take some sorting! I ended up with three spare lambs by night-time. My son would lend a hand after school, by filling up the sheep cratches and feeding, which was a big help.

The winter did not help my late reseeding, so I walked it in April to see if it was any good. The grasses were there but the clover was very poor. I saw Nelson Love one day and told him about it and he wanted to have a look, so I took him up. He agreed to send me twenty kilos of clover blend and told me to mix it with the fertiliser and sow it like that. It took really well, and by the autumn was very good; it has never been ploughed since. I never stopped grazing it as Nelson said the sheep would push the seed in with their feet.

We did another twenty-two acres that summer, the very wet ten acres at the top and twelve acres by an old cottage site called Ysgairlas that had been a workman's cottage a long time ago. This was a good piece of land that was growing bracken. The year before I had baled 450 bales of fern for litter. I got the wet piece in first, though it took a bit of work as it was mostly peat. The other piece was all good soil, but both came well. These fields had to be fenced with gates to the road. I used oak out of the wood and I spent a day cutting posts and putting them in place ready. I had Kinsey to come with the digger to put them all in one day and those posts are still there now, over thirty years later.

CHAPTER 18

Better times

AT THAT TIME we were lucky – Margaret Thatcher was in power and she helped farming to improve and increase productivity with good grants. We also received the Hill Livestock Compensatory Allowance payment on sheep and cattle, which came in March and helped to pay the feed and fertiliser bill. On the back of this we increased sheep numbers at Ty'n y Cwm to 1,000 because we had hill grazing for 900 at Ty'n y Cwm, and 500 at Aberannell, so there was plenty of scope. Winter was hard on Ty'n y Cwm as the land rose to 1,400 feet at the top gate, so we started sending some ewes away for the hardest part of the winter between December and 1 March. We were now keeping 400 Speckle hogs and selling the best 100 as yearlings in the Speckle sale at Builth. They would go off in October and come back on 1 April. We were also keeping 250 hardy Speckles at Ty'n y Cwm and they went away as well. This cost money, but it was well worth it as they grew better down on lower ground. The Ty'n y Cwm hogs went straight to the hill on their return.

I was also renting some land at Cribarth from Dilwyn Williams, and about that time another farm came up to rent on the estate and it was advertised in the *Brecon & Radnor*. The following week, I spoke to the new young agent. It was about 120 acres, with some good dry land top side of the road going through the farm, and some poorer land at the bottom side of the road. I asked was there much interest and he said no. The next-door farmer at the end of the road

Better times

wanted to rent the forty acres of dry land but wouldn't take the rest. The agent said he was not prepared to split it and asked what I was prepared to pay. I told him £5,000 and he said 'Alright, you've got the farm.' It worked out as it was less than a mile down the road. The next time I met the farmer at the end of the road he was mad, but I told him he was given the same chance as me. I thought he was going to thump me as he was pretty wild! He was not a good neighbour for a while afterwards, but I managed it OK. As usual, there were no pens on the place – how they managed their stock, I don't know. I put up some simple pens with a race for drenching and we were away. Taking that farm was the last straw for someone, as I started receiving poison pen letters. They were very abusive for a while, though I never took any notice of that sort of thing – they were wasting ink.

That autumn we sold a good bunch of mules and the Speckle yearlings in Builth. They were making a good income, so we looked to keep it going. Before the sales there was a lot of work preparing the sheep. The mules had to be neck-trimmed and dipped with colour dip to make them all look the same. Two days before the sale we would draw them into lots of thirty, and mark each pen with a different mark so it was easy for penning at the sale. You had to get it right because they were inspected and they were quite strict. It was more work trimming the Speckles, as they were yearlings and took more holding. I trimmed 300 mules one day and 100 yearlings, plus ninety smaller mules the next day. We looked forward to the sales for two reasons – to make some cash and to make room for the lambs left behind!

On the first day of August the Ty'n y Cwm ewes would go out to the hill, so getting them ready was a lot of work, what with weaning the lambs at the same time, but once the ewes had gone out we had plenty of room to fatten lambs.

When the small shows started, we started showing ponies. My daughter Amanda wanted a pony and Gwyn Edwards had got a mare. So we bought it and put her to a good cob stallion belonging to Roy, Tregare. She had very good foals. At the beginning of August I broke the foal in to lead and, once done, we tried our luck in the shows and had a few good wins! I also bought two mountain mares off Gwyn – he was losing the land he kept them on – and we showed these as well. It made a nice change from farm work.

At this time I was also getting involved with the NFU and enjoyed the meetings. When I was at the Pant, Rhulen, it was like Martin Luther King's 'I have a dream' speech. I dreamt of farming 1,000 ewes. We were now farming 2,000, and if things stayed as they were, we should be alright. I never put Llwyngwrgan into the FHDS scheme; there wasn't much hedging to do on it so it just farmed easily. Some of the land on the bottom needed ploughing but it was a SSI so no ploughing was allowed.

We were now making silage on the best land at the bottom of Aberannell and storing it in the pit up in the wood. The second year we had very good silage and came second in the Grassland Silage competition. The cows did well that year too. I had bought a bull that year from Penbont, Upper Chapel. He was a nice bull that sired a good type of calf, but in the summer he started to get a bit nasty. I wasn't happy, as we had footpaths going everywhere through our land. One day when we were bringing them down the road from Celsau, Gordon Williams came up behind us in his lorry and remarked what a good lot of calves the bull had got. I told him it was a pity I'd got to sell him and I told him why. Gordon said he might have a customer, and later he came and bought the bull for his son. They used him for two years, but he went very nasty so they sold him on.

During the winter I bought another bull ready for the next season. He was said to be from an easy calving herd, but he was a disaster! His calves were big and rough-boned and the cows could not birth the calves. He ruined eleven out of forty-five cows and the calves that survived were not very good. I was going to sell him, but he was as thin as a rake, so I had the vet to look at him and it turned out he had cancer. We had him put down and claimed on the insurance – it was a good job I had him insured.

By now our children were growing up. Amanda was sixteen and Michael was fifteen, so they were useful around the farm at busy times. Margaret was working every day.

That summer I did the last piece of ground by the lake at Ty'n y Cwm. It was awkward, with rocky patches and very steep land either side that went down to the top end of the lake and down to the wood on the other side, and flat on the top. But fair play to Dai Kinsey, he ploughed every bit possible. It was a hot summer, and I sowed the twenty-five acres at the end of July. The seed never germinated until 10 September, when it finally rained. Luckily it was a good autumn, and it came up well, but it wasn't grazed much as it was too young.

That winter I bought a universal four-wheel drive tractor and loader to make it easier to feed up in the wood. I was also catching foxes and you could get a good price on the skins by sending them down to Devon to R W Colledick. The best dog skins were worth £18.50 to £22, and vixens £15 to £17.50. Most winters I caught seventeen or eighteen, so I made a bit of pocket money and kept fit by going around the snares to see if I had caught a fox, as I didn't have a farm bike then.

At this time we were sending ewes and Speckle hogs on tack to west Wales. They did well down there, but it was a long way to go to do anything with them, because you needed to take hurdles to make a pen and a race to work with. We had one place where they grazed down to the sea. It was a bad place if the weather was bad because there was a big drop into the sea; you had to have a good dog. We had hogs by the coast for one year – it was a cattle farm in the summer and took tack hogs in the winter. They didn't do very well there, and I couldn't see why that was but we never used it again. It is often more to do with the man than the farm, some only want the cash and don't worry about looking after the sheep.

CHAPTER 19

A visit to France

IN 1984 I was voted in as vice-chairman of Brecon and Radnor NFU, which was a great honour for me. That autumn the branch organised a trip. George Hughes was chairman and decided we should go to EEC headquarters in Strasbourg. Garfield Phillips, the county secretary, said he would organise it and about mid-October we flew from Gatwick. It was my first flight, so I was a bit nervous but soon got used to it. We arrived at Charles de Gaulle airport and had a meal before setting off on a coach to Strasbourg. When we arrived we were shown into the headquarters and met Beata Brookes, the MEP for North Wales. We were told we could go up to the balcony to watch what was going on. We were going up a narrow set of stairs and, halfway up, we could hear an English voice booming – it was Henry Plumb, who was speaking about the miners' strike which was taking place. He was a farmer who had gone into politics as a leader of the NFU, and later became active in the Conservative Party and was elected as a Member of the European Parliament.

When speaking in the chamber, there was a traffic light system – when the light was green you could speak, when your time was nearly up it turned to orange, and when it was red you had to stop, Henry carried on way after the red light and the chairman had to speak to him. We watched for a while, then went down into the lobby and met Beata Brookes again who asked us who we wanted to see. We said Henry Plumb and the Irish minister. There was a seated area at the side being protected by a little Frenchman. She asked him if we could sit there and he said 'non'. She disappeared for a minute and came back with a big chap about six feet two who told the Frenchman to get out of the way, undid a gold-coloured rope and let us in. Five minutes later Henry and the Irish minister came. We had a chat with them about the problems in farming at that time, including Irish beef, which was being imported and undercutting our prices. It was a quarter-of-an-hour well spent.

When we left we got on the coach and headed back to a small town called Obernai for our first night. It was in the middle of the Alsace wine region and was very quiet, with not a bird in sight. They shot any birds because they ate the young grapes. We slept in chalets at the top of a bank and went down to the café for supper and breakfast. When we were having supper, there was a choir from the Swiss border singing in the next room. After we finished our food, some of us crept into where they were singing. It was very good – they were mostly ladies with a row of men at the back. They sang 'I Love to Go a-Wandering', with the men yodelling in the background. I've never heard it sung better. I can yodel a bit, so I joined in and they loved it!

We found a pub and had a few drinks, then we went back up to our chalet and got ready for bed. I was sharing with Gwilym, Erwgilfach. As soon as we got into bed, the window opened and within five minutes everyone was in our room. Gwilym started telling jokes and didn't stop for two hours! I put one in now and again while he was thinking of a new one. A couple of old boys fell asleep on our bed and we had to wake them up to send them back to their rooms.

Our bus was parked outside and the coach driver told us that it was safe to leave things on the bus. But, when we went out in the morning, some boys had got in through a small window and stolen a lot of our stuff. I lost a good camera and a bottle of vintage Grouse whisky and someone else lost a lot of money out of his coat. We were not very happy, I can tell you.

After breakfast the next morning we started back towards Paris. We had a number of visits arranged on the way back. One was to a sheep farm where they kept all of their sheep in a huge wooden shed that looked like a chapel from the outside. When we went in the shed, we could hardly breathe for the smell of ammonia. Jim Powell, Cwmfaerdy, was with me and he reckoned there would be some pneumonia problems – and he was right. In the last three pens they all had it. The farmer was rearing some sheep and buying others off people who couldn't fatten their lambs. He was feeding maize silage as well as cutting grass and bringing it in. It was a lovely farm, but we thought the sheep would be better out in the fields.

When we had finished we were taken to the local village hall for food where there were a lot of local ladies to serve us. We had a good feed and plenty of wine. After that we had to give a thank-you speech through a translator, which took a bit of getting used to. George and I shared this duty and soon mastered it.

We also visited a hill farm. The farmer was a young man of about thirty and he had one big black dog. He walked all day with the sheep and brought them back down at night and put them in a small field by his house. I asked him why he didn't let them roam, and he said that they would go into the wood and he wouldn't see them again as the wolves might get them. They were full of lice and couldn't stand still; they looked more like goats.

We visited a champagne producer and went down into the tunnels where the champagne was stored. There were thousands of bottles that had to be turned daily. Of course, we sampled some before leaving.

We called in several different farms, and one was owned by a British couple. It was a small farm and they kept

Southdown sheep. On the yard there were two new Renault tractors – they were getting a lot more grants than we were at home.

We also called at a Charolais cattle breeder who was very keen and doing a good job. His bull calves were left entire, so that the best went to breed at about £2,000. The ones that were left he sold to fat for £1,300 to £1,400. The best heifers were sold for breeding, the rest went to fat. All the cattle had horns and the two bulls he kept were monsters.

We arrived back in Paris on the Friday afternoon and that night we were invited to a reception with the British ambassador at the embassy. We got there by taxi and were met by the man himself. I have never seen so much drink poured out on a table in my life – anything you wanted, from whisky to wine, brandy, port – it was all there so everyone had a good night.

We decided to walk back to our hotel as it didn't seem far, but after we had walked a long way I said I would ask for directions when we met somebody. Around the corner came this lovely girl, very tall and slim, so I asked her how far we were from the St Anne Hotel. She said to me 'Do you want love?' I stuttered a bit and said no; I just wanted to find our hotel. She said we weren't far away, just go up to the end of the road and turn left, it was on the right. Then she said again 'Are you sure you don't want love?' By then all the boys were listening, so I had my leg pulled I can tell you, but we did find our hotel.

Garfield Phillips informed us that he had arranged for me and George to go to Rungis market the next morning at 6.30am, and would wake us at 6am to get ready. We took a taxi – he was a mad driver on the back roads; the wheels were rattling over cobbled stones, it was a hell of a ride! We did get there safely, and the market was an eye-opener, with every sort of meat and produce from quail's eggs to huge carcasses of horses and cows.

A visit to France

We went down to the lamb section and were shown the Welsh Quality Lamb stand. The French guide said the quality of Welsh lamb was not what they wanted. He complained that they were being sent lambs they couldn't sell at home – from great big, over-fat lambs to thin, small lambs. We were very disappointed. We went on to the Scottish stand, which was from Lockerbie, and they had a great show, all weighing eighteen to twenty kilos and good meat. The French man said 'This is what we need here.'

Nearby we came to a huge heap of wild boar, just paunched like rabbits, and then came to an egg section with eggs of every colour and size. The quail's eggs were the smallest and there was also dressed quail for sale.

We were then invited to have breakfast in the café where they had a section for the workers and another for the buyers. In that part, the Meat and Livestock Commission were showing a promotional film, demonstrating British lambs of different breeds. We were given a champagne breakfast and enjoyed it very much. The MLC man was a good chap called David Palmer – he was killed in a car crash two years later.

We spent the rest of that day sightseeing around Paris, and in the evening went out to see the city by night. We passed the Folies Bergère but didn't go in because they were charging £40 each on the door! Next morning, we were back on the plane home. When we arrived over London we couldn't land and had to fly around three or four times before we could do so. It was a bit hairy, but we did land OK, and arrived back having enjoyed a profitable week seeing a lot of the country.

The next morning I had to send two cows away after they had gorged on acorns. Robin Davies and my wife Margaret had done their best to treat them, but they had not responded.

During the winter I dif a lot of hedge-laying teaching for the Agricultural Training Board and Builth college, as well as fencing, gate-hanging and shearing in the summer. I was also getting more and more involved with Young Farmers.

The following spring, with the top of Ty'n y Cwm all reseeded, there was room for more sheep. So we upped the numbers again and started using more fertiliser because it wasn't dear at £97 per tonne. I ordered twenty tonnes in small bags and got the lorry up the new road to the top and dumped it off in heaps. I had to lift them into the spinner, which meant two hard days, but we had grass. If Mr Beynon had come back, he would not believe how it had improved. The following spring we marked over 700 lambs up there and a good bunch of mules down in the shed.

CHAPTER 20

Farming film stars

THE ROYAL WELSH Show shearing section was getting better, and we were attracting more shearers from New Zealand. David Fagan was the star then and would win most years. When I was commentating I was lucky to have him in the finals, as he made it a very exciting competition with times getting faster every year. I was now also commentating at the Three Counties Show and the West Midlands Show, plus many smaller shows, including Llysfasi, which was a very good show then. I also did some judging for a change. I was asked to go and commentate in the Great Yorkshire Show but I turned them down as I had enough to do already.

In 1986 I was voted chairman of Brecon and Radnor NFU. It was an honour to represent my fellow farmers and I did it to the best of my ability. That meant attending meetings around the counties, speaking to members about the problems facing farming. I enjoyed meeting people and made many friends. This also meant I went to London to the NFU conferences. If you wanted to speak in the hall you were given a microphone, but if there were four or five waiting, by the time it came to your turn to speak, the people before had often already said what you wanted to say, so you had to think of something else – not easy on the spot in front of hundreds of people. You were also asked to do television interviews, and on several occasions we had TV cameras arrive at Aberannell and Ty'n y Cwm to film. Nothing clears your mind better than being in front of a TV camera on your own, with someone firing questions at you – but I mostly managed alright.

During the winter I was asked by the BBC if they could make a *Play for Today* film called *Z for Zachariah* at Ty'n y Cwm, because it was the perfect location. They had travelled around Wales looking, and when they came from Abergwesyn they saw Ty'n y Cwm in its own valley at the bottom. The film was set in a valley which had escaped the fall-out of a nuclear war. Anthony Andrews starred as a man who finds a young woman there, struggling to keep the farm going. She was played by Pippa Hinchley, who was a very young, unknown actress at the time. I agreed to the filming but had to get permission off the estate to do it. They said it was alright, provided they made something out of it and agreed that with the film-makers. We had a payment for the inconvenience, too.

We were lambing when they were filming, so it was a problem as they had big vans parked on the yard. Very often, when I got there in the morning, my road up to the shed was blocked with cars, and I would have to find the owners to move them. I was also involved in advising them, and had to teach the actress to milk a cow. They also used my Ford 3000 tractor.

One day, they came to Aberannell wanting chickens, a cockerel and a hen with chicks. Luckily it was a Monday, so on Wednesday morning two of them came with me to Hereford market. I thought the sale started at 10am but in fact it was 9.30am, so a lot had already been sold. There was a fine cockerel that had already been sold, so I asked the BBC man how much I could spend. He said 'Just buy them.' I paid £10 for the cockerel and we then found a hen with nine chicks for another £10. We still needed a dozen hens, but we found some we liked and bought them in the auction for £2 each, so we were all sorted. While we were in Hereford the

BBC carpenters were making a chicken run, so when we got back, in they went.

The hens had started laying, and the BBC boys were in there the next morning collecting eggs for breakfast – they loved them.

We then needed a cow to calve, so I borrowed one from my son-in-law at Maesllech, as she was near to calving. A few days later my BBC man said he wanted the cow to calve the next afternoon! I told him it didn't work like that; she would calve when she was ready. And that would probably be in the middle of the night!

I was watching the cow very carefully, and when I checked on her at about 10pm I could see she was ready to calve by morning. As I came onto the road I met one of the crew going back to the Metropole where they were staying, and I told her to be ready for a call in the night. When I went back about 4.30am the cow was just starting, so I went and rang the Met and told the crew to get there as soon as possible. I got everything ready, and they started arriving.

I had already shown the young lady starring in the film a ewe lambing, and when she arrived I told her that a calf was 100 times bigger so that she was prepared. By then the camera was set up and running. The calf was coming out really well, and when the head was out I told her to pull the front legs and, fair play, although she had never seen a calving before, she soon learnt how. She pulled the legs and got the calf halfway, but couldn't get him all the way out, so I said to the crew that I would have to step in or the calf would die. I went in and got the calf, which had got stuck at the hips. At least we had a live calf.

The crew had never seen a calf being born either, so they were very excited about it. By then it was 7am and the cook arrived, so we all had bacon, eggs and champagne to celebrate the live calf!

The next morning Pippa Hinchley had to milk the cow. I

had taken her to a local farm, where they still milked a cow for their own use, to learn how to milk. She soon picked it up. I did wonder how the cow would behave, but she managed OK. While this was all happening we were still busy lambing.

I was now involved with many different committees, and at one time was chairman of eight, which was very time-consuming but I enjoyed it. I was also very involved with the Royal Welsh Show. Our shearing there was reaching a very high standard, as good as anywhere in the world. I was now travelling all over Wales commentating, and during the summer covered many hundreds of miles. In the winter I was having to go to hedging matches and the National to support the Welsh styles. I came close to winning it up near Leicester according to the chairman, who rang me on the Sunday after. He had been taking photos and told me I missed out by half a point. The judge was from the Midlands, and their style is a problem for Welsh hedgers using dead wood in the hedge. If you go back to the time when there was no wire netting, that hedge would be dead, with the sheep grazing, if you didn't use some dead wood to cover your live wood. It's different in areas between two fields of corn or on dairy farms with barbed wire.

Sheep numbers in Wales were now high because of the government's ewe premium, which made people chase numbers. We were now wintering 1,400 sheep near Barry, in south Wales, and they did well in that area. We had 400 Speckle hogs on one farm, 450 ewes near the airport at Rhoose, 350 ewes near Llancarfan on a dairy farm, and 200 on a very good farm just off Five Mile Lane with a Mr Evans. He was a good farmer and grew excellent clover leys. We also had 350 ewes near Whitland, so we were lightly stocked at home through December and January up until 25 February. The hogs came home on 1 April. We also had 160 hardy Speckle hogs near Monmouth.

We spent a lot of time on the road when they had to come

home and had a big haulage bill, but it also meant we had a lot of clean land to lamb on, which makes a big difference. We had one bad year when the sheep sales were poor. The Beulah sale had been good for years, but that year the trade was down from £75 the year before to £35 each for yearlings, and several people didn't sell. My first and second pens made £37 and I sold them, but the third pen only made £34 and the fourth £33, so I took them home. It was no money for yearling ewes, but there were too many sheep being kept by farmers. The next year, numbers in the sale were back and the trade was better again – when you have rent to pay, you have to sell something.

CHAPTER 21

On the march in Brussels

My work with the NFU seemed to grow. One night I had a call asking me to go to Brussels to join a march against cuts in prices proposed in the Global Agricultural Trading and Common Agricultural policies, which would affect farmers all over the world. I said I would go along with Mia Lewis, from Penybont, and a Mr Davies from the other side of Llandovery. We had to get to Dover to meet a bus to take us to Brussels. Mia picked me up at 5am, and we then picked up Mr Davies and drove to Dover. The bus taking us was full of people from all over the country. We got on the ferry and had some food, ready for a long journey. We had two drivers and crossed two borders where our passports had to be inspected. One was guarded by armed soldiers who held us up and were very fussy. Anyway, we reached Brussels about 12.30pm.

The NFU office staff in Brussels met us with bread rolls, cheese, ham and cans of beer, which set us up nicely for what we had to do. We parked at the top of the town with many other buses from all over Europe, and assembled to march behind the French, Germans, Spanish and Italians. We paraded down the main street, which was lined on one side with tall, glass-fronted buildings. The French and Germans were firing firecrackers at the windows; the noise was terrible as they went down the street. They broke everything that could be broken, including small trees and lightbulbs. They bent lamp-posts to the ground, then pulled up all the flowers and threw them on the road,

On the march in Brussels

When we got to the bottom of the street we turned left, heading back up to our car park about 200 yards up that road. We passed an entrance to some important buildings guarded by soldiers standing shoulder-to-shoulder with loaded rifles. The idiots in front were firing firecrackers that bounced off their helmets. I was never so glad to get back to the safety of our bus. It was just like being in a war. The headlines in the Brussels papers were 'Farmers do a million pounds worth of damage in Brussels'. I thought they were mad.

We got back to Dover and started the drive home, and stopped at the Severn Bridge for some coffee and food. Mia said he was knackered and asked could I drive from there, so I drove on home. We dropped Mr Davies off in Llandovery and got back home about quarter-to-five and I invited Mia in for a cup of coffee. We agreed we would never go to Brussels again, and off he went home. We had been there and back in twenty-four hours.

On one occasion I was asked to go to London to lobby MPs over milk quotas. We went by train from Newport and made our way to Parliament to meet as many MPs as we could, to try to stop milk quotas being introduced. The first one we met was Cardiganshire Liberal MP Geraint Howells, who was very helpful in getting us meetings all day with different members to explain to them what it would mean to lose the Milk Marketing Board. By 5pm we had done a fair job and decided to go out and have some food. Later we found a bar nearby and went and had a few drinks. There were NFU members from all over the country there, so we had a bit of fun. Then Garfield Phillips suggested we might as well stay and see the debate, as the Minister of Agriculture, Fisheries and Food, Michael Jopling, was speaking about 8pm. So, we had a few more drinks and went back to parliament to watch the debate. When we got there the Minister of Defence was talking. He spoke for half an hour and could have covered what was of any importance in five minutes. Instead, he went around in circles, so Joplin didn't start speaking until ten to nine. However, he was the same and we learned very little.

We caught the last train to Newport. It was a very cold night and there were problems with points freezing, so we took a long time getting back. Garfield dropped me off in Three Cocks and headed home. When I came onto the straight bit of road at Tregaer, a police car came and put his flashers on for me to stop. The officer asked me if I was late or early! I said 'I'm late, I should have been home hours ago, we're busy lambing.' He asked where I had been, and I told him I had been to London to lobby MPs about milk quotas and if he needed that verified he could ring Mr Garfield Phillips from the NFU, as he was with me. He said 'Alright, you had better go home to your lambs.' I was pleased not to have been asked to blow into the breathalyser, as I might have been over the limit. I got home at a quarter to three, changed my clothes and went out see the lambs. Gwyn Edwards had been seeing to them while I was away.

I was also asked to go on a legal committee that met once a month in Nant Ddu near Merthyr to discuss insurance claims. They were having problems solving these on behalf of farmers. This was an interesting job and as I was now valuing livestock for the NFU, I could offer expert knowledge of my own.

CHAPTER 22

Wedding bells

ON 23 MAY 1987 our daughter Amanda got married to Philip Thomas, Maesllech, in Beulah church. It poured with rain the whole day, but we managed. All the photos were taken in the Lake Hotel and the day went well. They bought a house at the far end of the village.

During the next winter my son Michael and his girlfriend Julie went on a long holiday to Australia and New Zealand, so I was left in charge. Luckily it wasn't a bad winter.

The next year we had to think about building a bungalow, as Michael and Julie were talking about getting married in 1990. I asked the estate for a site next to the village and was told I could have one for £5,000, so I said that was OK as it was near to all the services, and those things are what cost the money. We got local builder Graham Davies to give us an estimate of the cost. Michael was willing to do some of the labouring to save money.

We got an architect to come and have a look and my wife helped to plan it, so we had a good idea of what we wanted. The site was on a slope and the bungalow was set back, higher up, which worked well. Cliff Owen did a fine job of digging the site out, and we were ready to build a three-bedroom bungalow with a garage on the end.

Once everything was passed and ready, we started building. It was a lot of work, but Graham had a good workman and soon it started to take shape. Michael helped with some of the labouring jobs, and when the roof went on I gave a hand getting the tiles up. Once the roof was on, Graham could do outside work when dry and go inside when it was wet. After the bungalow was finished there was still a lot of work painting, varnishing doors and skirting boards. We were able to do a lot of that in the evenings and on wet days, and we were able to get it finished by the time of the wedding.

Michael and Julie got married in Talgarth church on 1 September 1990, and we had a very enjoyable day.

Originally, we had built the bungalow for them, but Michael was keen to stay in the farmhouse, so Margaret and I moved into the bungalow before the wedding as it would save us moving again when we retired. We had the yard up to the garage tarmacked for parking – it all cost money, but it had to be done.

After the wedding we took Michael on as a partner in the farm business, so he had more of a say in what happened. Our stock was running pretty high to take advantage of subsidies, and during those years we were buying or leasing sheep quota every year to keep our payments up. Most years we spent about £5,000 and, in the end, we owned all our quota.

We were still running forty-five cows and they were paying alright. We had to change a few cows every year as my original cows were now getting old. I usually bought my cows in Sennybridge. I like the hardy, black, white-faced heifers off the Beacons – they milked well for us. You can't beat a black cow x Charolais bull, they grow well and sell well.

I bought some odd cows as well. One heifer I bought wasn't very big – she was in calf and was not a good-looker, but she was stuck at £170. I thought, 'She's cheap', so I put a bid in and bought her for £180. She calved a month later and had a good bull calf – in fact, she always had bull calves. By the autumn her calf would be bigger than she was. She was the best cow I ever owned – her last four calves averaged £485.

Wedding bells

I also bought a Simmental heifer that had lived a hard life and was as thin as a rake. She also had a black bull calf with her. Heifers that day were making £750 to £800 and she was stuck at £420. It was a gamble, but I put in a bid and bought her for £430 along with two others for £750. When I took her home Michael said 'What the hell did you buy that for?' I told him if she was healthy, she would improve. We gave them all a dose of fluke and worm and put them to grass. I've never seen an animal improve like that heifer, and by the autumn she was looking healthy and rearing her calf well. She turned out to be a good cow and we had her for years.

The sheep sales were going well. After the drop, they came back up and we sold our Speckle yearlings, getting our best pen up to £97 one year. The mules were also still selling well.

I was asked to go to Welshpool to inspect at their sale. This is not a job for the faint-hearted because sometimes you must take sheep out for some fault or other. Every sheep in the sale is inspected to keep the standards up and you always have somebody trying to pull a fast one, but we usually got the job done without a row. It's not a nice job telling someone they must take their sheep out. I knew what it meant, as mine were inspected in the Builth sale. We were selling 400 – ten pens of forty in the sale – and some of our last pens were twins, so not very big, but I made sure they were of the correct type and got them through.

One year we weren't placed too badly in the sale, so I hoped for good trade. When the first pen came in, a gentleman in the front put them in at £60 and they went up to £69. He put every pen in at £60 and never bought a pen. They averaged £64, which was good for 400 hogs. I had kept ninety small mules the year before and they averaged £89 as yearlings, so I was happy when the cheque came a week later. The mules came to £33,300, while the Mule Society and the auctioneers took £3,300 out for ten minutes' work. But that was a good year; we had some not so good as well.

Later that autumn we went on holiday to Sorrento in Italy. It's a very nice area and we had a good hotel and took some trips out and about. One day we got up early and went to Rome, which was a long ride on a bus. It parked up by the Colosseum where the Romans held their games – such as feeding prisoners to the lions! We looked round and I've never seen so many cats in my life; they were everywhere and being fed by little old ladies.

Our next visit to the Vatican was very interesting. We were shown around by a guide and security was very high. We then visited the Trevi Fountain which is a beautiful large, carved, marble fountain, and wandered around Rome. It had been a great day but it was soon time to get back on the bus; we would have been happy to spend a week in Rome.

The next day we looked around Sorrento. At about 11am we walked up a short street to a café for a cup of coffee and sat under a canopy outside. While we sat there a thunderstorm came from nowhere, and it poured down. In minutes the street was flooded and rats started appearing out of the drains. Margaret was scared of rats, so we waited for the storm to go by before moving on.

The following day we took a boat to Capri. What an expensive place! A room in the hotel was £1,600 per night then; what would it be now! The next day we spent on the beach in the sun, and the following day we went on a trip down the Amalfi Coast. We enjoyed the day in Amalfi, the church there was worth seeing. The week went very fast, and we were soon on our way home.

After getting back home there was plenty to do getting the sheep down off the hill ready for the ram. We would bring ours down at the same time as the rest of the boys on the hill, as our hill is a big area and I relied on them to help us. There are usually about ten to twelve people, some on bikes

Wedding bells

or ponies, along with about twenty-five dogs. We usually brought 2,500 sheep in on that day and had to race them out for each owner. Once ours were out, we would take them out of the way – ours were usually about 900. They would give us a hand to get them down, as we had to get through the forestry and it could be a nightmare without some extra hands.

Once down, we would part the yearling ewes out and put them over to Penybanc for the winter. They would be drenched for fluke and worm before they went. We would part the poorest ewes out of the rest to go to the Texel or Blue Leicester ram, and the best we would put up on the top of Ty'n y Cwm for the winter.

We usually got it all done by night. There would always be a few strays returning from other farmers when they gathered theirs.

We didn't put the rams on till 25 October; we wouldn't want to lamb until the end of March because it's cold up there.

The following May we went down to the Bath & West Show to watch the World Shearing Championships. I sat about halfway back in the tent where it was being held, and after an hour a member of the Bath & West Committee sidled along the seats and asked if I was Tom Evans from Wales. He asked if I would help the young commentator in the afternoon when the big finals were held, as he was new to the job and didn't know the top shearers. I said I would, and I enjoyed the job, but at the same time I was thinking we could do a better job of the World Championships at the Royal Welsh!

We had a shearing meeting later after that year's Royal Welsh Show, and I asked what we could do to get the World Championships there. The others said it was tied to the Bath & West, Masterton, in New Zealand, and a town in Australia. I argued that wasn't a world championship; it should be able to go anywhere in the world. Anyway, Peter Guthrie suggested maybe the only way was to go to the Five Nations meeting in London at the Winter Show at Earl's Court. I said I would be happy to go, and Peter Guthrie agreed to go as well, so it was arranged in November.

CHAPTER 23

A victory for Welsh shearing

IN THE AUTUMN Marg and I went to Ayia Napa in Cyprus on holiday. We arrived at our hotel in the dark, but we woke in the morning to a wonderful view. We had an enjoyable week going up to the Troodos Mountain in the centre of the island and seeing a lot of the country on the way. We came down a different route and came close to the Green Line buffer zone between the Turkish and Greek areas of the island. We were not allowed to cross to Nicosia on the other side, although you could see it from the coach.

Ayia Napa is now a wild place, full of young people and it's not safe to walk the streets at night. Back then, it was lovely and quiet. The hotel was mostly full of Germans, and if you wanted a good place on the beach you had to get up early because they went down and pinched all the sunbeds themselves, but we managed alright. The food was the best you could have. One morning Margaret went into the bathroom, and she came out screaming. On the back of the door was the biggest spider I had ever seen, it could only just about sit on my hand! Its body was like a small mouse. I got a newspaper and took him outside and let him go – we never saw him again!

In November, Peter Guthrie and I went to the Five Nations shearing meeting at Earl's Court. On the way in we met the rest of the delegates and did a bit of lobbying, just to let them know what we had in mind. At the meeting Peter got our request put on the agenda. The secretary of the Bath & West Shearing Committee was allowed to speak first and spoke for twenty minutes on the history of shearing at the Bath & West and why they should keep it. He sat down and I asked the chairman if I could say a few words. I told them what they had now wasn't a World Championships, that any country should be able to bid for a World Championships, just as with the Olympics. If a country had the money and a committee to run it, they should have it. I explained I was not speaking only for the Royal Welsh, but also for the top shearers who were asking for it to be opened up. I only spoke for five minutes and sat down. We won the vote and were very pleased and went home happy.

When we had our Christmas meeting for the Royal Welsh Show, we reported what had happened and it was decided to hold a World Championships at the show in 1994, and to write to the Golden Shears – which runs the World Championships – to tell them that we were holding it with or without their permission. They didn't like that a lot, but accepted our request. We then had a lot of work to do getting sponsors and money to run it, but it all came together well. We also had to find an extra 1,000 sheep to shear. We used 3,000-plus every year in any case, but for the World Championships an extra 1,000 were required.

One night a member of the NFU rang to ask me to be spokesman for them at Builth market, as the NFU were calling for all markets to be stopped from selling for one day because the lamb trade was dropping every week. The price was 83p per kilo a week earlier and was forecast to be 78p that week. I was asked because I didn't sell lambs in Builth, so they couldn't boycott me – I sold all my lambs deadweight.

It was a tough job telling farmers who brought their lambs to market on that day not to sell, but most agreed with what we were doing. This was happening in markets all over Wales.

A victory for Welsh shearing

As you can imagine, the buyers and the auctioneers were not happy. When all the sheep were penned in the market and the auctioneer got up to sell, I got up and said my piece. I told them that farmers were unwilling to accept prices dropping every week, and if they kept on going down we would be giving our lambs away. We were asking farmers not to sell on that day as a protest, and that all markets were doing the same. Fair play, only four or five pens were sold, but we were undermined by a farmer buyer who went round afterwards and bought a load of lambs for his slaughterhouse. It did have the effect of stabilising the trade, however. At that time lambs needed to fetch £1 per kilo to be worth selling.

About this time we also went on holiday to Portugal and stayed in Faro, its old capital. The hotel was alright, but the old town wasn't much to admire. The whole place stank of grilled sardines, which were cooked outside. They must have lived on them, as the harbour was full of them. We went on a bus tour one day and saw that the farming was very old-fashioned, with corn cut with a binder and stooked up in the fields. Smallholdings were still using the horse to plough, although I imagine it has changed a lot since then.

The following winter Wye Valley Grassland Society had invited Oliver Walston, a farmer from the east of England, to come and speak. He told us he was making so much money on corn that he had changed all his tractors to the biggest he could buy, and the same with his combine, cars and pick-ups. He was convinced that the EEC was the future, but there had to be a level playing field. I stopped him and told him I had just come back from Portugal where they were so far behind the UK. They were still cutting corn with a binder and ploughing with a horse. He started to fret a bit and couldn't answer that, so most of us had a good laugh. He stayed in the Lion Hotel that night and left us a phone bill for £38, so he wasn't very popular.

The preparations for the World Championships were

going well. We had found digs for all the teams and judges from all over the world. We also met the TV people. Coverage wasn't very good then, although these days it's very good. When the time for the championships arrived we were ready. Alan Davies and I were in for a hard week, but it went OK with the help of Raymond Powell in the mornings. He was busy with his shearing in the Sheeptacular some of the time. One of our sponsors was Deosan, and their head man was Tony Bristow. One morning, we had to go on the radio and he introduced me as 'the voice of Welsh shearing'. He was a likeable chap and got me and Alan into the sponsors bar for a drink afterwards.

The world champion that year was Alan McDonald from New Zealand. He now manages a large farm back home with his son who is also starting to shear. We had a lot of good Welsh shearers coming through at that time – John T L Davies, Nicky Beynon, John Pugh, and Arwyn Jones were learning a lot off the New Zealand boys and getting faster every year.

After that show the shearing went from strength to strength and the quality improved every show. Over the next couple of years we had to raise money for Breconshire's Feature Year in 1997. The Hon. Shân Legge-Bourke was chosen to be president, and a very good one she was. I was made chairman of the Builth fundraising committee, so we set a programme of things we wanted to do.

We wanted a three choirs' concert, so I had to arrange that with two soloists, Aled Edwards from Tregaron, who was a tenor, and Eleanor Davies from Aberedw, as well as the three choirs from Builth, Talgarth and Brecon. We also organised a bar with Ray Davies in charge, and with the efforts of some good helpers it went well. We held it on the showground and raised £2,500.

Then I had the idea of getting a singing group. So we had the Bachelors, an Irish group who were very popular at one

A victory for Welsh shearing

time. We attracted a good crowd and raised £3,500 including the bar.

In the end we raised £15,000 in our area, and along with a generous grant from the council we raised over £200,000, which was a lot of money. I had been on the Breconshire Advisory Committee for a few years and sometime later I was asked to take on the chairman's job, and have been chairman ever since.

In 1998 Ireland was given the World Championships, which were held in Gorey. One day I had a phone call asking me to commentate over there, which was an honour for me, but I knew it would be a hard job. Wyn Jones from Montgomery was asked as well, so when the time came I travelled over with Brian Jones, Mike Evans and Matthew Price, who I used to shear with. We were staying on a farm about ten miles outside of Gorey. The farmer was an interesting man and we learned a lot about Irish farming.

On the first day their own competitions were held, just to get everything going. By that night they realised they hadn't got anywhere for the scorers to be. I came by later and they were putting a garden shed up at the end of the stage. One chap said to me 'Jesus, we will be OK tomorrow.' I said to him 'Tomorrow never comes, mate', and he laughed at that. They had built a good place to commentate from; I had a good view of everything. We then started the World Championships and Wyn and I took it in turns to commentate through the heats. That night we had to parade the teams and ended up at two pubs at the far end of town where we downed a fair bit of Guinness. By the Saturday night Wyn had no voice left, so I was going to have a hard day on the Sunday when all the finals were held.

I was lucky that a New Zealand chap came over and asked if I wanted a hand as he commentated back home. In fact, I had just met the top commentator in the world, Phil O'Shaughnessy. He was brilliant. On Sunday morning he said to me to let him go on stage and bring them on. He would sit at the far end from me and come in to add this or that, and it worked very well. He was so good, he made my job a bit easier.

That afternoon there were six sheep finals, with each shearer shearing twenty sheep each, one after the other. That takes some doing, even with the best in the world. The shearing was being held in a large tent, and by the end the sides were all full with spectators and it was twenty or thirty deep outside. The noise was serious!

We got it done and on Monday morning we went to catch the ferry. Although it was early in June, it was very rough with a storm forecast. Anyway, we got on the catamaran and the Irish captain announced that this would be the last one that day, as it was too rough. We were jumping from wave to wave, and it was a rough crossing. One of the Welsh judges was at the bar, holding on with one hand and with a pint of Guinness in the other. Michael Evans had taken some Kwells and slept the whole way over. He woke up just as we came into port and asked what the crossing was like. We said 'Rough'.

CHAPTER 24

Another country

FARMING WAS NOT so bad and prices were OK. Our cattle were selling alright; we had a good bull and he was doing a good job. We were selling our weaned calves to Radnor Store Stock. Mervyn Morris was the agent and brought a chap called Green to see the calves. We had our usual forty-five calves. I had sorted the bull calves out on the yard, the heifers were in the shed. Mr Green was a keen old boy from the east of England. He walked through the bull calves and asked what weight they were. I told him roughly and he said I was over the top a bit – well, you have to try anyway! He said he would give me £485 each for the twenty-two there. There was a £100 sub on them I hadn't claimed on. For the twenty-three heifers, he said £385 with no sub on them. We shook hands, and he said the lorry would come for them the next day. He told me that I had a good bull, so he came out back and had a look and said 'No wonder you've got good calves!' It was a good way to do business with no hassle.

Once the calves had gone the cows would live easily. At the end of January I was feeling a bit low, and said to Margaret I needed a holiday before lambing. She asked me where I wanted to go. I answered New Zealand. We booked flights, arranged a hire car and left around 17 February and flew into Christchurch twenty-four hours later.

We went to pick up our car and found it was a Honda Civic automatic. Marg said she wouldn't drive as she'd never driven an automatic. Neither had I, but I said I'd have a go and we loaded up and away we went. I soon got used to it. We headed down the main road, and after about forty-five minutes decided to have a cup of coffee in a roadside café. We parked and went in and had coffee and a cake and, as we walked back to the car, a campervan screeched to a stop and reversed back. The window opened and the driver shouted 'Tom Evans, what the hell are you doing here?' It was David Morgan, Middlewood, and his family who were over there for a holiday. His brother farms there, so we had a good chat. He told us we needed to get off the main road to see New Zealand properly and he gave us a route to follow. We saw some good sheep country. We went as far as Genevieve and stopped there for the night in a very clean bed and breakfast, then we walked down to town to have some food and a bottle of New Zealand wine. On the way back up to our digs, we went through a park, sat down and both fell asleep – jet lag had kicked in. After a while, we woke up and continued to our digs. We felt a lot better the next day.

We crossed over to the west side and ended up in Queenstown. We had an address there of a shearer who worked for Robin Lloyd in Painscastle. We found him, but he had no rooms. He gave us another name just down the road where there was one room left – it's a very popular place. The second night, we stayed with Dave Clark, as there was plenty to see around Queenstown. Above the town there's a winery with a cave cut out of the mountain to store the wine – it never changes temperature throughout the year. On the river there was a boat they call the Shotover Jet. You have to be mad to go on that because it speeds down the river and then comes back up just as fast. Also, further up there was bungee jumping off a high bridge into the river. You needed nerve for that or plenty of beer! We spent the day sightseeing and enjoyed it very much.

The next day we moved further down to Te Anau and

Another country

stayed the night, and went out on the Milford Sound on a boat trip where we met more Welsh people.

We then went over to Dunedin on the east coast where we found digs with a couple from Wales who had lived there for years. We crossed back over next day and went to the west coast to visit the glaciers at Mount Fox. It was fascinating to see two valleys still full of ice from the Ice Age. I had visited them first in 2000, and visiting them again in 2017 I could not believe how much they had retreated up the valley in those years. They had gone back about 400 yards. In a few years they will not be such a draw for sightseers, though there were a lot of people visiting them then.

It was harder to find digs up in this area, but we managed. The driving was good, with no traffic, it was a pleasure. One day we were driving through a big valley which was beef cow country, mainly Hereford. It had been a very dry summer and they were giving the cows big bales of hay. We came by one herd of 150 cows on a big field by the road, and the men were coming out onto the road, so I stopped for a chat. I told him we'd had a wet summer back home, with seventy to 100 inches of rain. They said they would let us win the rugby for half of that, which made us laugh.

At one point we passed some hill land that was grazing Merino sheep. You could hardly see them as they were the same colour as the grass.

At the end of the first week we took the car to the ferry to North Island and left it there. We had another car arranged for the other end. We had a full crossing, with thick mist all the way, but I was told it's very often like that. We arrived in Wellington, which is one of the biggest towns in New Zealand, and had booked a room in advance in a new hotel as it was on special offer. It was good, so we enjoyed spending the night there. We went downtown and found what looked like a good place to eat, but the food wasn't very good. I had lamb shank and it was the worst lamb shank I ever had, so not a good advert for New Zealand lamb!

The next morning, we got up and headed for Hunterville Station, as we had arranged to be picked up to meet a chap I met in Ireland at the World Championships. He had said if we came over to give him a ring, and he would pick us up, take us out to his farm and let us have his second car for a few days. They met us at the station which was way out in the wilds, but it was lovely country. We stayed the night, and the next morning he had got the car ready. It was a big old Datsun, and again an automatic, but I was used to that by now. He had filled it up with diesel and cleaned the bird muck off the front window, so off we went. We arranged to meet them in Wellington the night we were leaving for home.

We made our way up as far as Godfrey Bowen's farm, watched the Sheeptacular and went on a tour of his farm. We saw kiwi fruit growing and different breeds of sheep. We then visited the boiling mud pools higher up, and also a geyser. We also went up into the hills and stayed a night on a hill farm. This farmer kept 2,000 ewes, 2,000 deer and 170 Simmental cows all on his own, apart from the help of a young man who lived in a farm cottage. When the farmer needed help, this chap would go to assist with the deer or the sheep, the rest of the time he was a shearer. When we arrived, a lady told us she'd had a Welshman staying a month or so back. I looked in the visitors' book and I found the name of Haydn Jones, Y Ddole. When I saw him next, I told him where we had stayed and he couldn't believe it!

We left the next morning and made our way down to Masterton to the Golden Shears competitions and booked into our digs. The first day's shearing was done, so I decided to go and find where the hall was and get some exercise. When I arrived in the loading area, the Open shearers were walking across the yard. Digger Balme saw me and asked 'What the hell are you doing here?!' He had been to the Royal Welsh to shear. I told him I'd come to see how they did it here. I had a look round and went back to the hotel.

Next day I went into the hall and was disappointed with the size of the place, but the shearing was good. I met Phil O'Shaughnessy, who had helped me in Ireland. He interviewed me every day, when he had a chance, asking me about the Royal Welsh and our competition. One lunchtime he asked me about my commentating and the show. A couple of people came round from the far side of the hall and asked if I'd been commentating in Ireland. They said they had recognised the voice, so we had a good chat. David Fagan was the star; he had won Masterton all through the Nineties and then again in 2000. What a man he was when he was at the top! They had told me Masterton was the best competition in the world. I came away disappointed as it was not as good as the Royal Welsh in my eyes.

On the Sunday morning we headed back to Wellington to meet our plane for home. We met our friends at the airport and thanked them very much, then got on the plane to Auckland and said goodbye to New Zealand.

If I had gone out there at the age of eighteen or twenty, I would still be there. I loved the country.

When we arrived back we had been away seventeen days, including flights. The weather had been OK while we had been away, so my son had managed alright – but next, we had to prepare for lambing.

CHAPTER 25

Stolen goods

AT THE NEXT shearing meeting in 2000 we were asked what the Royal Welsh Show could do in addition to the shearing. We decided to have Phil O'Shaughnessy over to commentate as we had several young commentators starting at the show. It would do them good to see how he did it, to improve their style. The idea worked well. He stayed with us and we compared notes every night. The young commentators learned a lot from him.

That year also we started having trouble with BSE in the cows. We had seven cases pretty quickly and the trade went bad. The last cow to have it was the Simmental cow I had bought cheap. We were paid compensation and I had double what I had paid for her – she was now a fine cow. The last calves I sold in Sennybridge were all Charolais heifers, 300 kilos, and they made £152. I told Mike Tompkinson, the auctioneer, that these were last cattle he would sell for me. He told me 'Don't say that, stick to – it will soon get better.' I told him I couldn't wait as my son hated cattle, so I told him what I thought they were worth and he agreed. That night I made five or six phone calls and, within a week, all the cows were gone. I had one lame cow left and the bull, and they went soon after.

In the autumn of 2000 the Welsh Ploughing & Hedging Championships were held in the village of Sarn near Newtown. I decided to have a go at the hedging and managed to win the Welsh championship again in a good class.

Stolen goods

I was now travelling to London one day per month, as I was on the NFU Hill Farming Committee and met farmers from all over England and Wales. William Jenkins, from the south Wales Valleys, was chairman and did a good job. Like me, William was a tenant farmer, and when it gets hard tenant farmers feel it first. The committee met at 10am and it wasn't possible to get there on time, so we decided to go up the night before and stay in the President's Hotel. This meant getting the train in Abergavenny at 3pm to get there by 6.30pm to 7pm. I don't like the Underground, so I got a taxi from Paddington which cost about £10, but we were able to claim it on expenses.

On one trip I stopped in Abergavenny to get some paracetamol pills, as I get a headache on the train. When I went back to the Land Rover a boy of about eighteen was looking through the side window. Seeing me, he moved away about thirty yards and tied his shoelace, but I had a good look at him anyway. I got in and drove down the road, then turned back up to the station and parked in front of the pub door. I got my bag, locked the Land Rover and, as I went to go into the station, that boy and another one came out. It worried me all the way to London and back. When I came back, sure enough, my Land Rover was gone. I was fuming! A girl I knew from a local Young Farmers' Club came out and asked what was wrong. She offered me her phone to contact Marg to fetch me – she couldn't believe it. We went to the police station to report it, but they didn't want to know as I was parked on railway land. My Land Rover was twelve months old, had done 12,000 miles and was worth about £16,500.

The next day I phoned British Transport Police – they were hopeless. All they were worried about was whether it was insured. They rang about a week later asking were there any personal items in the Land Rover that I could claim for. I told them that it would already be resprayed and on sale

159

in some garage by now. Anyway, we had an old Land Rover from Likes until our new one came in – you get new for old with NFU insurance. When the new one came in we were away again, but this one was a Friday night job. She was hopeless and cost me £2,000 per year in parts and labour – whenever you had some work to do she would break down.

CHAPTER 26

Foot and mouth

FOOT AND MOUTH disease hit us in 2002 – this was a disaster for everybody and being on the Hill Farming Committee we had to sort a lot of problems out. In the autumn the trade on lambs was bad and there were no outlets for hill lambs, so they had to be moved on. At one meeting we were told that the government wanted to give 30p per kilo to dispose of hill lambs. I said that was totally not on, they should be £15, or at least £10. I was backed by a member from north Wales and a minimum price of £10 was passed. That was what we were paid and thousands of lambs went on that scheme.

A welfare scheme was also introduced in the spring and I told Michael that it was for people in trouble. However, when I went to Builth the following week, I heard that some of the biggest farmers were using the scheme, so I went home and said to him 'We've got to get on this.' We had bought some land next door and the bottom land was a bit wet and peaty. We had a bunch of Texel cross lambs with some old Ty'n y Cwm ewes down on the wettest piece, and we applied for a licence to get onto the scheme and we were accepted. Then we had to get the vet to confirm that they needed to go. When we loaded them on the lorry, the driver said it was a sin to see them go – they would be needed in the autumn. I told him the way the trade was, they were better going that way.

The next move was to mark the lambs on the top of Ty'n y Cwm. As we were doing this, we marked the older ewes to sell in the autumn and we also marked the wether lambs differently to the ewe lambs. A couple of days later, we went up and sorted out any old ewes with a wether lamb into a separate field. After two days we had a good bunch, so we got the vet again and he agreed for them to go, which also helped a lot.

We had some Ty'n y Cwm hogs near Monmouth and we were not allowed to move them home. We were lucky not to have them killed, but they survived and the next autumn they were still there. We had to get some tups down to serve the ewes, so we set out at 5am with three tups with their legs tied and covered in straw in the back of the Land Rover, and arrived there with no problems. The farmer there was also a magistrate, but understood what had to be done.

At the end of August I came back to the yard and there was a big car parked up there. It was Mr Beynon from Gower who had bought thousands of mule ewe lambs from Builth over the years. He would buy 500 in Welshpool and the same in Builth, then he would sell 500 yearlings in both places and it paid him well over the years. He had been to a Mule Society meeting in Builth that was discussing whether there was to be a sale or not that year – it had been decided not to, because of all the rules surrounding livestock movements. He asked how many mule ewe lambs I had to sell. I told him that about 400 would be ready. He said that's what he wanted. Because of the rules, three artic lorry-loads would suit him fine. I asked about the price, and he said they would have to be a lot less than the previous year, but I told him, in that case, I was keeping them because yearlings would be short next year. He didn't like this but had to accept it and off he went.

We lost our keep for hogs near Cardiff, so we had to find keep. I was talking to my brother-in-law, and he said he had a relative near Hereford that had been badly affected with foot and mouth and had plenty of keep. When I rang him he said it was no problem, and asked how many I would have.

Foot and mouth

I told him about 1,100 including the mules, and he said he could manage the lot. This was great as it's always a job to find good keep.

Under the rules there was one week when we could move them down. We had to wash the lorry in Hereford market in between loads, which made it harder but we got it done. We lost all our farms around Barry, so we had to find keep for our ewes, but we were lucky and found a good farm near Whitland and also a good farm near Monmouth.

Farmers still had a lot of sale ewes, old ewes and rams on their farms because there were no marts. I was asked to go down to Cardiff to meet Carwyn Jones, who was the Secretary for Agriculture and Rural Development in the Welsh Government at that time, to discuss a scheme to get these sheep off the farms before winter. Four representatives from the NFU and four from the FUW were allowed to meet him. We pressed him about the problems, and that with winter coming something had to be done. In the end he agreed to put a scheme in place, so we won the day.

Foot and mouth was a nightmare. It was very sinister because it would jump 100 miles sometimes, so it was very worrying. A man rang me from Libanus near Brecon and said he'd been told foot and mouth would arrive in Libanus in a month. A month to the day later, he rang me to say he was being taken out the next day as foot and mouth had been confirmed on the next-door farm. He was very angry. I told him to make them pay the best price he could get, and he did.

A friend of mine near Builth was also taken out. I knew there was something wrong because I could smell the smoke from the fire where his livestock was being burned. I rang him and left a message telling him to get the best price possible.

Many sheep on tack were taken out and some valuers were being very hard. I valued for the NFU, so I helped a few people to get better prices. In one case I improved the price for one farmer from £90 to £150 per head – that was for in-lamb Speckled ewes scanned at 150 per cent. It was a much-improved end to a bad job. When well-bred sheep are killed off, a farmer loses years of breeding. The loss is a lot greater than just money.

Tony Blair and the Labour Party have a lot to answer for between foot and mouth disease, the financial crash and the war in Iraq. He should be locked up, not going around the world getting £1,000 for a speech.

After the cows went we decided to do some hedging in the winter as we had got time. The Tir Gofal scheme was going well, with good grants on hedging, so there was plenty of work. Over the border in England they had a good scheme as well. In the next seven or eight years we did twenty-three miles of hedging, completing about three miles every winter. In our best winter we did four miles, and a lot of this was in Herefordshire. We set off early in the morning, arrived at the hedge by daylight and finished about 4pm to drive home to feed the sheep that we still had. We wouldn't come home without hedging more than 100 metres per day, rain or shine. Many of the hedges hadn't been touched in sixty years, so you needed a good chainsaw.

CHAPTER 27

Meeting the Queen

IN SEPTEMBER 2002 my landlord, Mr Mervyn Bourdillon, was very ill with cancer. I decided to go and see him, bought a bottle of whisky for him and went up to the house. His son Patrick opened the door and I asked how things were. He told me his father was very ill. Mrs Bourdillon came downstairs and said 'Oh Tom, it's you. I think he's sleeping', but she went back and asked me to go on up. He was very pleased to see me. On the table by the bed I noticed his whisky bottle had about three inches left, so I said to him it was a good job I had come with a bottle. He asked his wife to go and get two glasses and some water, telling her 'Tom and I will finish that bottle.' She poured the whisky and water and chatted about the estate and the Young Farmers Club – he was president of the local YFC and loved the local club. We also talked about Beulah Show, which was on the following Saturday. He asked how I was getting on with Ty'n y Cwm; he was always interested in any improvements we made.

After a while I could see he was getting tired, so I said my goodbyes and left. The next day, at about lunchtime, I met Patrick on the road, and he told me that his father had died early in the morning. I could not believe it. I still miss him today – he was a friend as much as a landlord.

Sometime later I was asked to be president of Beulah YFC, a post I accepted and which I still hold today.

When foot and mouth went away, things returned to

165

normal except for the red tape of the six-day movement restrictions, double-tagging and forms to fill in to go to market. The trade returned and by the autumn things were better.

We had a good lambing and, with the 106 acres we bought next door, we managed alright. The hogs at Hereford wintered well, and during the first week in May we went down to shear them. The 400 mules had grown well and sheared even better. They were cutting the heaviest wool I've ever rolled – with thirty fleeces in a wool bag, I couldn't move them on my own. We sheared all the hogs and left the mules down there until the beginning of August, along with 100 Speckles, as we wanted them looking their best for selling; the rest we took home. The Ty'n y Cwm hogs went out to the hill. We also had the yearlings home from Monmouth for the first time in eighteen months.

In May 2002 Shân Legge-Bourke phoned one night asking me if I would help her in her role as Lord Lieutenant of Powys; the Queen was to visit Dolau during the fiftieth year of her reign. She wanted me to find as many breeds of sheep as possible that were kept in Radnorshire, also mountain ponies, goats and a sheepdog with pups. I spent a week on the phone every night and everybody was very happy to bring along their stock.

On the day we had pens ready and I was there to help pen up and keep some sort of order. When everything was ready, Shân Legge-Bourke came to see me and asked me to stay by the pens so that I could meet the Queen. She arrived by train and there was a huge crowd who had come to watch, including hundreds of schoolchildren. After being welcomed and being shown around the tents, she came over to the pens and asked me what the big sheep in the back pen was. I told her it was a Jacob, one of the oldest sheep breeds, and she was very interested to hear about all the different breeds. Then she went to look at the Welsh sheepdog pups and the

Meeting the Queen

mountain ponies. She loved horses, of course, and had a good look at them. The Duke of Edinburgh was also there, but I didn't meet him. I then helped everybody to load up and went to look in the tents and met Shân Legge-Bourke who was very pleased with how it had gone.

As a thank-you for the work done during Shân Legge-Bourke presidency year and the Queen's visit, we were invited to a Garden Party at Buckingham Palace. It was held seven days after the bombing of a bus and station in London. Marg didn't want to go; she was afraid there might be more trouble but I convinced her that security would be tight, so we went by train from Abergavenny and took a taxi across London. Richard Livsey, our MP, had rung two days earlier and asked us to the House of Commons for a tour and to have lunch with him. Mia Lewis and his wife were also meeting him, along with another couple I'd not met before, so we enjoyed our tour and had lunch in the Commons. Richard Livsey was a good MP for Brecon and Radnor.

About this time I had a letter from Charles Woosnam giving me notice on the top of Ty'n y Cwm – land I had drained, improved and fenced. His idea was to plant it all with oak and ash trees to get a big grant for fifteen years' maintenance. I was mad and my landlord Patrick was away on holiday till the end of the week, so I had to wait till the Friday to see him.

When I went up to Llwynmadoc, Patrick asked me into his office and I explained and showed him the letter. He couldn't believe it, he knew nothing about it! He told me he had no intention of planting that land but he wouldn't mind planting eighteen acres of steep land above the lake, so I agreed to help him fence it.

I also told Patrick that my ten-year tenancy was up, and I'd like to have another ten years. I said I'd offer a bit more rent and we agreed on that. Patrick told me to be at Woosnam's office at 4pm on Monday, and it would be ready to sign. When

I went Woosnam was very angry. He threw the agreement across the desk and said 'Sign this.' He didn't speak to me for a long time after that, but I had what I wanted.

CHAPTER 28

Time to slow down

AT THE AGE of sixty-five I retired from all my committees on the NFU as I wanted someone younger to take over, I had done my bit. That done, I joined Builth choir as I like singing and persuaded two more from Beulah to join. I enjoyed going to sing in London where we sang at the opera house. We also sang twice at the Royal Albert Hall during the 'Night of 1,000 Voices', which is some show. We also went over to the Czech Republic, to Český Krumlov, and stayed in Prague. We have sung to raise funds in many halls in mid Wales and over the border.

About this time my son decided he wanted to farm on his own. I was not very happy about this after spending years building up the business, but remembering what happened with my father, I decided to split the farms. Mike wanted Aberannell and the land we had bought off Fronrhydd. At the same time Cefngardis, which joined Aberannell, came up for rent so it worked well. It made his part about 350 acres. I was keeping Ty'n y Cwm, Penbank and twenty-seven acres at Llwyngwrgan, so I had about 340 acres, which was enough at my age. Michael took over the Beulah Speckle flock and I kept the hill flock on Ty'n y Cwm, which suited me.

Mike decided to build a new house on the land we owned. So to get planning permission we had to put that in his name. He then built two sheds and a store shed over a few years. It's now a good holding and he's able to lamb 1,200 ewes indoors. He has all Blue Faced Leicester rams producing mule ewe lambs for Builth sale – we have sold mules at that sale since it first started.

In 2008 I was asked to be president of the Beulah Speckled Face Sheep Society. I had been chairman sometime before. It was an honour to do the job and I enjoyed it very much.

Being back farming on my own, I had to change the sheep a bit to produce bigger lambs. So I bought six Lleyn rams to put on the yearlings at Penbank, and on the older ewes I bought four Talybont Welsh rams to improve the ewes in size and weight, and this worked well.

I was using Texel rams on some older ewes to get some lambs gone early, which was also successful. I could lamb them in the shed at Ty'n y Cwm – I was lambing 1,200 ewes on my own. It made it a busy month, but once lambing was over you could manage, no problem. The yearlings at Penbank took a bit of handling, but with small fields there was always a corner to catch the ones you couldn't lamb. I didn't get many, but yearlings are pretty fit and I put pens in good spots so that I could chase them in with a farm bike.

By the end of lambing I was pretty fit. I was selling lambs to Dilwyn Thomas, Llwynbaedd, and it worked well, but he always wanted them at 7.30am and that meant being up at 5am to get them ready. Later, Kelvin Tucker in Newbridge started buying lambs, which suited me better as he would pick up the lambs after lunchtime so that I had the morning to get them ready. With the improved lambs, I would move about 100 at a time, and at weaning I sold everything that was fat. On one occasion I went through the lambs at weaning, and they had done well. I found 430 which were ready, so I rang Mr Tucker and he said he'd move them in three lots. I told him how many there were in each place, and he shifted the lot that week. I weaned the rest and the ewes went to the hill out of the way. This was a lot of work but worth it – with all 1,200 ewes on the hill, it made room for the lambs to fatten.

Time to slow down

At weaning I took out about 300 old ewes. I gave them a good place and they were soon ready to sell. I always took out the bad udders and ewes with anything wrong with their udders, and sold them as they wouldn't have any milk to feed their lambs. After that, I had people wanting ewes to run for a year or two. Some wanted fifty, some wanted 100 and they soon went – I never charged a fortune.

Once, when I was marking lambs on the top of Ty'n y Cwm on my own, I was busy and getting through pretty well when a hailstorm came and covered the sheep and ground with an inch of hail. But I was determined to finish, so I kept on as best as I could out of the hailstorm. A chap with a foreign accent came walking down the Black Trail which runs over Ty'n y Cwm. I told him to head below the pens where he could go without going through the sheep. When I had finished the sheep and let them go, it was hailing again. I went to open the gates down onto the land above the lake, and this chap was hiding under some rocks, sheltering from the hailstones. I stopped and asked him where he was going and he told me he was walking the trail four times that year to see all the seasons. He was a Dutch student, and he told me that when he got to Beulah he was supposed to go and see Tom Evans. I said 'You are looking at him!' He had been told by a chap in Yorkshire that I had a 'cue', which is a shoe for a cow that the drovers used to shoe all the cattle with when they drove them to London. I told him to walk on down to Beulah and I would meet him there. We had a long chat about droving and then on he went, after he had taken a photo of the 'cue'.

CHAPTER 29

More royal connections

IN APRIL 2012 Shân Legge-Bourke rang again asking if I could help with a 'Diamonds in the Park' event to celebrate the Queen's sixty years on the throne. She needed some stock and horses to show the Queen, plus shearing and wool-spinning demonstrations. I asked two hand-shearers and some spinners from Brecon to attend. My cousin, Olwen Jones, came to knit the wool. We also had ponies again and a lot of small animals. This event was being held at Glanusk on 27 April, so we went down the night before to get it all ready. Shân let us stay in the house so that we were ready for the next morning.

It was a wet night and so was the day itself. When the Queen and Duke arrived they went on different routes. The Queen was taken around in a Range Rover and came to our side. When she reached me she got out and smiled when she saw me – she had remembered me from meeting me twice before. I said 'Good afternoon Ma'am, sorry about this Welsh weather', and she replied she thought it was pretty general. I showed her the shearing and told her about wool and she talked to the spinners and to Olwen before moving on to the next stand. When I went back to thank the shearers and spinners, one lady came out and hugged me and told me she had expected to talk to the Queen. Shân Legge-Bourke had worked very hard to make the day a success and was very pleased with the result.

After the day was over we drove straight down to Torquay for the YFC National Drama finals, and stayed in a caravan

More royal connections

above the town. It was a rough first night with gale force winds. We thought the caravan was going to blow over and Marg was dressed by 7am just in case, but it died down a bit by 9.30am and we had a good weekend there.

We have enjoyed many trips to the finals, visiting Blackpool as well as Torquay, with Builth YFC representing Brecknock and doing very well.

The next autumn I bought another three Talybont rams in the NSA sale. I fancied them a lot, they were just the type I like with some brown in the back of the neck and no horns. I don't like horns as they are always getting caught in the fences. Those three rams cost me £100 each, the most I'd ever paid for Welsh rams, but they did the job well and had heavy lambs and a good type of ewe lamb. I kept the best for replacements to the flock, so after a few years the flock improved a lot.

I was chairman of the Royal Welsh Brecknock Advisory Committee and in 2012 John Davies, Pentre, Merthyr Cynog, was going to be president of the Royal Welsh. The year before that was spent fundraising for Brecknock Feature County Year, so it was my job to lead the members in fundraising. One event we arranged was a Rhydian Roberts concert, which was a great success and we laid on our own bar which made a good sum. Ray Davies again organised that for us and I served behind the bar.

We also had a Grassland Day up at Pentre, which John organised, with silage machines on the go and big bales from different companies. It was held on 8 June, but the weather was more like 8 December! It was so cold and wet. I was stewarding on one stand and by lunchtime I was absolutely frozen. When my relief man arrived, I had a ride back to the yard and went into the bar where Ray was barman again and asked him to give me a brandy as I was frozen! He handed me the bottle and said 'Help yourself.'

After some food I went back to the bar. We had arranged

for some sheep to be available for Young Farmers to have a go at shearing, but because the day was so bad the shed was full of people. Instead, I asked Greg Evans, who was to have trained the boys to shear a few sheep, to demonstrate shearing and I would commentate and tell the story of shearing and wool. Greg sheared about half the sheep, then stopped and had a drink before finishing the rest. We attracted a good crowd and got the job done, but the weather did not improve at all.

In the evening we had a concert in the shed and a lot of people came, so I was on the bar. Some of the Irishmen who were over doing the demonstrations were there, and I've never seen men drink like them! They were downing pints of beer with a port wine chasers, and by 8.30pm we had to send someone to Brecon for some more port wine – they had drunk the lot! There were about seven or eight of them, and by the end they were very drunk – all except one. He had been drinking along with the rest, but he looked sober. How he wasn't drunk we don't know! At 7.30am the next morning they were loading bales onto lorries, happy as bees.

Ray and I closed the bar at 2.30am as we were both knackered by then. The bar paid well and the whole day made a good sum even with the bad weather. A lot of it came from renting plots to these companies to demonstrate their machines. The crowd could have been much bigger if it had been a good day, but you can't do anything about the weather, it's just a matter of luck. In the end we raised £270,000 for Breconshire, so John was pleased and his year as president went well.

We had applied to the Golden Shears World Committee to hold the World Shearing Championships at the Royal Welsh Show in 2010, so again we had to find sponsors and TV coverage. We also sourced an extra 1,000 sheep to cover all the classes. We rented every bed in Llanwrtyd Wells for the teams and judges, and we arranged buses to get them

More royal connections

back and forth. The show went well, with huge crowds on the Wednesday breaking all records – 10,000 more people than normal – so we had a super crowd at the shearing. Commentating was hard work because of the noise they made cheering and shouting so much! All the commentators were knackered by the Wednesday night, but it had gone very well. The winner was New Zealander Cam Ferguson, who has since retired from shearing because of ill-health. In all, twenty-seven teams from all over the world took part.

On 17 November 2010 I was invited to an evening reception at Windsor Castle, so we booked a hotel room in Little Windsor. Marg was not invited so she wasn't very happy, but she stayed in our hotel and had a nice meal.

I took a taxi to the castle, found the entrance and went into the lobby where I was told when to go in. On entering, I was announced and had to walk forward to meet the Queen, the Duke of Edinburgh, and Princess Anne. Once this was done, we went into the Great Hall where there was a crowd of 700 people. I walked into the middle and was offered a glass of wine and there I bumped into Henry Plumb, now Baron Plumb, who I had met in Strasbourg. He remembered me and we had a good chat. Somehow, we got talking about hedging and he told me he could hedge as his uncle had taught him when he was fourteen. He said 'You never forget.'

I met a few people I knew from around the country, some I'd met at NFU meetings. One chap with a big beard I'd seen on telly, and I told him I'd seen him on television with Adam Henson, who presents *Countryfile* on the BBC. He said I was right and that he was actually waiting for him. He'd known him since he was seven years old and had sold Longhorn cattle to his father. He told me to wait and he would introduce me. Adam visits the Three Counties Show and always comes to see us at the shearing for a chat. The food was now on its way – it was all titbits, but they kept filling your glass with wine, so I did alright. When it was time to leave I got a taxi

175

back to the hotel in Little Windsor. By then it was 9.30pm and I was very hungry, so I had some food and a pint and I felt better. The next morning we had a look around Windsor and then made our way home.

I was now asked to be president of Brecknock YFC and I accepted. For me, a Radnorshire man, to be given the job was an honour. I enjoyed being with Young Farmers, and during my year I visited every club at some point and was a guest speaker at several dinners. I also went back to Radnorshire at their seventieth anniversary dinners in Rhosgoch and Edw Valley, my old club. I am still on two committees at Brecknock YFC.

On the farming side, I had stopped using the Texel ram and also the Lleyn and was using all Welsh rams to make lambing easier. I had nearly all Welsh-bred ewes, so I now had a lot of spare Welsh ewe lambs, but soon found ready buyers for them at home.

CHAPTER 30

Shearing Down Under

AT THE 2011 August Welsh shearing meeting, I was asked to be team manager for the Welsh Shearing team in Masterton, New Zealand. Ray Davies was on the World Shearing Committee and had to go as well, so he suggested that we should go to Australia on the way for two weeks. He had a good friend in Port Augusta in South Australia who wanted us to stay there. We would go on to New Zealand for five weeks, which would include the World Championships. My son-in-law Phil agreed to take over while I was away, and Ray organised the trip. Dai Fairclough came with us as well.

We flew out around 15 January and arrived in Australia where we were picked up by our host Pete Smith. We had about seventy miles to travel to his home where we met his wife Sandra, who made us feel at home. It was very hot there compared to Wales!

Pete was a shearing contractor who ran a gang in that area, shearing large numbers of Merino sheep. One day he took us out to the big Outback farms to look at some of these flocks. We drove fifty miles up the road before turning off onto a dirt road, and went another fifty miles to get to the first farm. We had coffee with the owners who said they farmed 1.2 million acres carrying 17,000 Merino sheep. Pete was in charge of sheared his flock.

After we left them we travelled another fifty miles to the next farm. They were great people. The owner had a four-seater plane and wanted to take us up but it was too windy. 'We'll have to have a barbie and see how it is after,' he said. I've never seen people eat meat like the Australians do. We had a large plate of meat, steak and chops half-an-inch thick, with just one bread roll. Ray and I put on a stone each in a fortnight. After the food he went out to the plane and came back and said he thought we could go up. We got in and took off and flew to a height of about 700 feet before levelling off. It was more windy up there than on the ground and he said we would have to turn as the plane wouldn't go on, so down and round he went. What a ride we had – bobbing up and down and from side to side, but it got better as we went round his farm. This holding was 1.4 million acres with another 17,000 Merino ewes, which Pete also sheared. We came down and I thanked him very much and moved on again another fifty miles to the next farm.

This chap was the brother of the one we just left, and his farm was about 1.5 million acres, also with another 17,000 Merino sheep, another shearing job for Pete. We had a look round his shearing shed and the yards. The main job on these farms was keeping the water tanks full for the sheep, most of the pumps were run by windmills.

We then drove another fifty miles to the next farm which was the driest of them all. It was higher up and was 1.6 million acres. We were now over 200 miles off the main road and still on dirt roads. This farm carried another 17,000 Merino ewes… and Pete sheared these as well! Merino wool is very valuable; the ewes cut six pounds of wool, the wethers nine pounds. Shearing them is hard work and the best shearers can shear about 200 per day. Shearing these flocks takes three weeks at each place with six shearers. They can't handle any more because the wool is graded off the sheep and baled up in 200-kilo bales, so the shearers stay on the farm where they all have living quarters. The owner of the last place was not at home, but he had told Pete to show us the shearing shed as it was an old one with wooden pens and gates, but it still worked well.

After a good day we got in the van and headed for home and took a shortcut down through a valley, driving along a dry river for part of it. We saw some wildlife such as kangaroos and ostriches, which live out in the wilds. Some of the ostriches had young ones, and it's the cock that rears the chicks. We saw one with nine, a good hatch. They are big birds and very fierce when they have young.

Next day we visited one of Pete's biggest customers, Bruce Nut. At that time he was farming 2.4 million acres with 76,000 ewes and 20,000 cattle. Nowadays he farms 5.4 million acres with 140,000 ewes and a lot of cattle. We arrived at his farm at about 10.30am and met him and his wife in their café. When shearing is over, they turn the shearing shed into a café and his daughter-in-law runs it to make extra money. At the end of the shed was a heap of wool bales. I asked what it was worth, and he said 'about a million dollars' – it was all Merino wool. As we had coffee and a look around, Ray was telling him about the Royal Welsh Show. He said he would like to see the show, and the following year he came over and stayed with Ray and had an enjoyable time.

Next, we went out to a farm called Willow Springs to visit the people Ray had taken the Welsh team to when he was team manager for the World Championships in Toowoomba, a few years earlier. This was their 'practice' farm, and they were so pleased to see Ray as he and the team had stayed a week there. On the way to the farm we climbed up a fair way and turned into a valley, and the number of kangaroos there had to be seen to be believed. There were three different sorts, some six feet high, others four feet high and some small hairy ones. I asked the farmer about it and he said he had to apply for a licence to shoot them for meat. We stayed the night there and had a barbie and some beer. The next morning we went further up the valley to his shearing shed and the shearers' quarters. This was very smart, and we had another barbie for breakfast outside. This was a 1.6 million-acre farm with 18,000 Merino sheep.

When times were bad they had to get a contractor to bulldoze a road around his farm, up hill and down dale, to make a track for cross-country drivers, as it takes most of the day to get round and he wanted to let his shearing quarters out for tourists to stay. They were very nice people, who looked after us well.

We moved down to the coast next and visited someone who used to shear for Pete but was now in the business of taking people out on his boat to an island to see the wildlife, whales and dolphins. He had built his house on the seashore, with living quarters for his customers to stay. We stayed there that night and had a barbie and more beer. His wife told us the first job in the morning was to clear the snakes off the lawn that went down to the seashore. They came in off the sand dunes, and she had a special snake-catching stick to take them back. We couldn't get out to the island because it was too windy, so next morning we moved on and went out back to the countryside again. At lunchtime we stopped for a drink in a small town and went into a pub, got our drinks and sat down. It was a Saturday, and in walked a lady of about twenty-three or twenty-four. She ordered a drink and, as she leaned on the bar, she looked over at us. Eventually she came over and said to Ray that she knew him – hadn't he been judging wool handling at the Royal Welsh? Ray said he had – she had been competing there. It's a small world! So she sat down and told us she was wool handling out in the wilds and had decided to come into town for a change.

Pete then took us up into the high country to see the wombats he had noticed when he was passing through to shear at a farm. Further on, we turned up a dirt road and drove for about a mile but never saw one. We could see the holes, but no wombats. Coming back we had a puncture, so we had to change a wheel. This done, we drove on about another mile and pulled into the smallest bar I've ever seen – it was only about six feet wide and you sat outside to drink. Pete knew the owner and told him we'd had a puncture. He said 'We'll mend the bugger now, Pete.' His wife was sat outside the door, and she looked very ill. Apparently, she had only just come out of hospital two or three earlier and had been in a coma for two or three weeks. Her husband was a hell of a joker, and when she came out of the coma he told her she had been in it for five years and he had remarried and had two kids. It's a wonder she never died of shock! He came back with the wheel repaired – it's not a good idea to be driving about without a spare here as you could be fifty miles away from the nearest settlement. He was quite a character. He told us a film crew who were filming just down the road came in for a drink one day, and he told them about all the snakes and spiders around that could kill you with one sting. They moved the next morning!

Next day a chap who works for Pete as a foreman on a second shearing gang rang. He asked if we would go and help him get some cows in off a mountain. Pete said we would and we went to meet him at his home in the bottom of a valley. The farm itself was at the top of the valley, so we went up and stopped at his building and pens. He got his motorbike out and we went up the valley to the hill where the cows were stranded. On the way I counted seven or eight dead cows. We were dropped off in different places to help move the cows down to the small lake where we would separate out what was wanted. Where I was dropped off there was a walnut tree just off the track, and also some of the biggest thistles I've ever seen. They were about fifteen to twenty feet high with stalks as thick as my leg and burrs on them the size of my hand. Good job we don't grow them like that here!

After a few minutes, down came about thirty cows and calves, all Herefords. We guided them to the lake and sorted out what we wanted and got them down to his pens. There were six big calves and a heifer and calf he wanted to sell, as she was too small. He had 150 Hereford cows and 2,000 Merino ewes. The six big calves had never been inside in their lives – and he was selling them the next day. I asked him why the cows had died. He said 'Brown snakes, mate!' He told us they came out at night looking for mice and voles, but if a cow grazes too near them she can get stung, and a half-ton Hereford cow will die in ten minutes. I asked how long you had if one bit you. He said 'Five minutes.'

We then had a barbie and some red wine from his own vineyard. Sitting eating our barbie a huge lizard, about a yard long, came up under the gate, then turned and went up a tree. At the back of the sheep pens there was an orange tree loaded with fruit, so I asked if they were for eating. 'Yes,' he said, 'have one.' I'd never tasted an orange straight off the tree before – it was very good.

The following day Pete suggested we go fishing, so we hooked the boat up and away we went. We were going to fish by a power station which discharged warm water into the sea and where it was possible to find some very rare fish. We went out into the bay at Port Augusta, and about halfway we stopped to catch some bait before carrying on to the power station wall. It was a bit rough there, so we moved out thirty yards where it was better. Pete baited two lines for us, and switched on the sonar so that we could track all the fish under the boat. I had a bite and started to pull my fish in but it escaped, went under the boat and took Ray's line. He pulled his line and had it well and truly hooked. When it came out of the water it was a yellow-tailed king fish, an attractive fish with a yellow stripe down its side and weighing about five pounds. Pete was keeping an eye on the weather and said we had to move as a storm was coming, so off we went at about thirty knots towards the shore. We had the fish for tea the next day, it was very tasty.

When we were off to meet the plane to New Zealand, Pete took us to a zoo to see what brown snakes looked like. They were horrible-looking things, not actually brown but dark grey, and they grew to ten or eleven feet long. Next door to where we were staying there was an overgrown building site that had just been cleared before we got there. They had killed two nine-foot long males and two eight-foot females. No wonder people have two doors to enter their homes – the outside one is metal gauze, which you can see through, and that is always closed.

We met the plane and arrived in New Zealand where it was much cooler. Most days in Australia it was about forty degrees celsius, but a dry heat. We picked up our car and headed out into the countryside where we needed to join the team in Gorey for the first competition. We were there for two days, and the team did well. The team consisted of machine-shearers Gareth Daniels and Richard Jones, hand-shearers Elvet Jackson and Gareth Owen, and the two wool handlers, Anita Jones and Gwenan Pawai – a good team I could be very proud of.

We all met up and talked about where they were competing, so that we would know where we were each weekend. During the week we visited people Ray had invited over to shear for him when he was a contractor. They were all pleased to see him and we would be asked to stay for the night – they wouldn't take no for an answer, they were great people. We stayed with Mark Dowling and family who farmed about 2,000 acres on some very dry mountain land where he kept Merino wethers for wool. He lambed his ewes on a large area of dry hill. It had been improved at some point, but the ley was old and tough. He was putting all his ewes there to lamb, and sixty Angus cross Hereford cows to calve. He only went around them once a week. If the season was normal he never lost many lambs, and it was rare to lose a calf. He kept no labour, his wife worked elsewhere and his daughter helped when she was at home. Ray had welcomed three of the Dowling family over to shear, and we visited brother Pete near Lake W naka.

We followed the team to all the different venues. There were Wales versus New Zealand Test matches where the team did well and they were certainly enjoying their stay. The wool handlers were having a hard time as the New Zealand handlers were very sharp. They would be doing it all year round, so our girls had to work hard to reach a semi-final and were pretty stressed. I told them not to panic too much; it would be easier in the championships – and it was.

The Saturday before the championships we were about twenty-five miles from Masterton in a small town called Paihia. We arrived around 9.30am and went into the town hall where the shearing was being held, and sat about halfway back to watch the competition. After a while the boss of the show, Koro Mullins, came over to say hello and asked me to commentate, as his men wanted to save their voices for the next week. I agreed and took over from a Maori lad who was doing a fair job of the junior heats.

I did an hour or so and the young Maori lad came back. I told him he could do a few heats as I needed to have a rest, so he carried on. I went back to the boys and sat down, but after about ten minutes a lady sat behind me tapped me on the shoulder and said 'Please go back on that microphone.' Koro Mullins also asked me to go back on, so I commentated for the rest of the day.

Our shearers didn't reach the final but sheared very well. The day before we had been out on a farm with the team to practise, and on the Sunday we moved to Masterton ready for the World Championships. Ray had a championship meeting on the Tuesday, so I went to the hall to watch the local competitions – that was a good place to see a lot of very good shearers who never travel over to Wales.

When it came to the hand shearing, I was asked to

commentate with George Graham from Ireland on all the heats and the semi-finals in the big tent across the road from the hall. The weather had been wet the week before, so they had to move the sheep to the edge of town because the valley they were in would be flooded, but they managed OK. In this tent it was so cold everybody had coats on, and the hand-shearers were shivering as they waited to shear. Elvet Jackson just missed out on the final, Gareth Owen came fifth, but it was won, as usual, by the South Africans. In the machine shearing, Richard Jones missed out on the final, but Gareth Daniels came fourth in the world – which was very good considering the company he was in. Gavin Mutch from Scotland sheared exceptionally well and was crowned World Champion. Our wool girls didn't reach the final but did well in the team event. We were looked after very well in Masterton, with free digs and free food at the hall, so we couldn't complain.

 We left Masterton on the Sunday and called to see Travis who was a shearer for Ray and was now managing 1,100 acres carrying 1,100 cattle all the year round. He would sell 200 fat then buy 200 store cattle back in. He had some of the best black-and-white-faced Angus cross Hereford bullocks you could wish to see. We stayed the night and made our way to the airport.

CHAPTER 31

A health scare

It was quite a change coming home after seven weeks away, but after the very cold weather in New Zealand it didn't feel too different. My son-in-law had done a good job looking after the sheep, so all I had to do was push some more feed in ready for lambing. I also had to get fitter after putting on a lot of weight on holiday!

Lambing was much easier with all the Welsh ewes, and by the end of it I was back as fit as ever. I was diagnosed with cancer later that year and had to go to Prince Charles Hospital to have a lump removed from my colon. After the operation my surgeon, Professor Haray, told me that he wouldn't need to see me for six weeks. But, six days later, he was on the phone asking me to go to see him the following Tuesday. When I went in to see him he asked if my wife was with me, and to bring her in. He said 'I have bad news, the lump we took off from your colon was full of cancer, and I want you to come in for scans to see if it has spread.' I went down the following Tuesday and spent the morning having every scan I needed. For the last one I had to take a barium meal, and was scanned for twenty minutes on this one. The man in charge said he couldn't find anything, but it could show up on the other scans.

A week later I met Professor Haray again and he said he was pleased to tell me they hadn't found anything else, so I was lucky. He did frighten me by adding that, to be absolutely sure, I could have my colon removed and be fitted with a bag.

A health scare

He sent me home with a colostomy bag to play around with, and to let him know whether I wanted it done. I went down the following week and told him I would gamble for a while. He said he didn't blame me, because if they had removed my colon it was clear I wouldn't be very pleased. I get checked every year and, so far, five years later, I'm still fine.

A few months later I woke up one Monday morning and my heart rate had doubled in speed overnight. I was very busy at the time and thought it would come down in time. On the Friday morning it was still high, so I went to see my daughter-in-law Julie, as she is a cardiac nurse. When I told her what had happened, she felt my pulse and told me to go to the hospital in Builth and she would meet me there. She put me on a heart machine and advised me to go to the doctor to increase my blood pressure tablets.

The doctor told me I needed to go to Merthyr for shock treatment on my heart. This was arranged for the following Tuesday – I was one of five to be done that morning. One was a lovely lady from the Valleys who'd had serious cancer trouble. She had got through that but then her heart rate had doubled. I had my pre-op and went in. I was anaesthetised for a short time and then they zapped me up with the same pads they use to restart the heart after a cardiac arrest. When I came round, my heart rate had come down from 110 to fifty beats a minute.

I was asked to go back in three weeks. When we arrived the lady from the Valleys was in the waiting room again. She said she wasn't too good, her heart rate had gone back up after a week. One old boy, who was about twenty stone, had stayed OK and so had I. Apparently, you can have that treatment five times, but after that you have an operation or a pacemaker. So far I've stayed well, and during that time I was still farming my 1,200 ewes and managed to keep going.

During the 2015 Royal Welsh Show, I was awarded the Point of Life Award. A few days before the show I had a phone call from some parliamentary offices in London who wanted to check that the information they had been given about me was correct. They asked about my involvement in Young Farmers, the Royal Welsh Show, and shearing commentating, about my involvement in hedge-laying competitions and the different committees I had sat on. They knew more about me than I did!

They explained that the Prime Minister was interested in our work in the countryside, and I said to Margaret I thought there was something funny about the call. Two days later a lady rang me back to tell me that David Cameron was coming to the show and he would present me with the Point of Life Award. He came to the shearing pavilion, where I met him, along with Margaret and the family. He gave me an envelope with a very nice letter in it, but there was no certificate and, so far, I've still not had it!

In the show I judged the Hill Breed groups. This is a lot of work as there are so many different breeds to judge. Anyway, I managed to do the job and, when I was going back to the shearing, I met some farmers who had agreed with my judging, so I was happy.

During the autumn I was asked to chair the All-Wales Ploughing and Hedging Championships to be held at Three Cocks in the autumn of 2016. This, I knew, would be a big job but I agreed because Builth Ploughing & Hedging Society was to be the guest committee. We were only a small committee, so we had to enlist the help of the local Vintage Society to help. The championships were to be held in September and we needed to find sponsors to pay the costs and prize money. All in all, we needed about £30,000, which took a lot of work. I spent a lot of hours on the phone. Many of the machinery firms had had a bad twelve months and hadn't got much spare money, but it all came together in the end.

At the same time I was planning to retire from farming, and my grandson Paul, Phil and Amanda's son, wanted to have a go, so I was busy selling surplus stock, machinery and my tractors. Paul was not going to take on Penbank, so I had to sell 300 ewes. I sold off the older ewes and 100 yearlings in Rhayader. I took 180 old brokers to Hereford and some old tups again to Rhayader. Paul took over at the end of September, but I had Penbank till the end of December, so what was left of my lambs went there.

On the day of the Ploughing & Hedging Championships we had about ninety tractors ploughing, about eight horse ploughmen, and we had a steam plough working on the other side of the road. We were lucky, it was a lovely day and a good crowd turned up and we took a lot of money on the gate. The night before, we held a 'Blessing of the Plough' service in the local church, and I read a passage from the Bible. We had food in Gwernyfed school and lots of people attended. On the Sunday I went down to help clean up and collect the numbers off the plots. Another job done – and we had made a good profit after all the hard work. My committee all worked well; landowner David James was a very good host and Peter Guthrie was great, supporting me all the way through.

CHAPTER 32

Back to New Zealand

IN NOVEMBER WE had our shearing meeting and were told New Zealand were holding the World Championships again in February, in Invercargill on South Island. Ray asked if I wanted to go and I said I would think about it. I discussed it with Marg and she said I should, as I didn't have any stock left. So I rang Ray and told him. Dai Fairclough was supposed to be coming as well, but when the time came he was not very well and had to cancel, so Ray and I flew out around 15 January and landed in Queenstown.

We flew via Australia but didn't stop off this time. We headed down to Invercargill and could not believe how the prices of everything had risen since we were there last. The first night, we tried to get some digs, but none were to be found as it was fruit-picking season, and loads of students were working picking cherries. New Zealand exports 90,000 tonnes of cherries a year to China.

We went through a mountain pass to another small town, Cromwell, which was also full of students. So we called in the tourist centre and the lady said she had one place. We went and found it and it looked good, so we took it and asked was there anywhere to get some food. She told us to go to a pub just down the road. We had food and a couple of pints before heading back and getting ready for bed.

The bathroom and toilet were up some steps in our digs, and in the middle of the night I went to the toilet and was

Back to New Zealand

sitting down when the door opened and a naked lady came in! She screamed and ran back to her room, but not before I noticed she had a nice little Brazilian! Next morning I told Ray about it and he had a good laugh. When we went for breakfast he asked the landlady how much extra the streaker was. She laughed and said 'Silly girl, why didn't she wrap a towel around herself?'

The next morning the hills around Queenstown were white with snow, they were having very poor weather for that time of year. We headed down to meet up with the Welsh team and stayed in a small town called Winton and got digs in a Travelodge-type of place. The next day was a Sunday, and when I got up about 8am and looked out of the window it was pouring with rain and blowing a gale. I told Ray to stay where he was. We finally got up at 10am and it was still the same. We found a café and had breakfast, then decided to go for a drive around the back roads to see the country. We came back at 2.30pm and it was still the same, so we dozed and watched TV till 7.30pm, then we went down the pub for food and a pint. It was still pouring down and some country roads were flooded.

Next morning we went up to Mount Lintern Farm, about fifty miles up a valley which used to be mining country. The farm was huge – 38,000 acres and carried 70,000 ewes, 3,300 Angus cows, 11,000 hogs for replacements and 700 heifers as replacements for the Angus herd. They were used to having visitors, so we went to the yard and a shepherd told us to have a look at the shearing shed, which had twelve stands and penning to cover 6,000 sheep. The outside pens were about ten acres, which after the rain the day before were covered in four inches of water. The shepherd said they had hoped to shear lambs that day, but that would have to wait till tomorrow. He was sure the pens would dry by then – they had 30,000 lambs to shear that week.

The following Monday we went back there as the world

teams were going there to practise. By then the shepherd told us they'd only got 4,200 lambs left to shear as they had shorn the rest the previous week. We had a good day watching the teams practise with some of the top world shearers who helped coach teams from other countries, including Welsh shearers. Alan McDonald, our world champion in 1994, was there helping and still a super shearer.

We stayed in Winton and again went to the pub for food and a pint. There was a big stock wagon outside and the driver, who was a big man, was in the pub having food. We got talking to him and asked how many lambs he could get on his wagon. It was his own lorry and he said he could load up 600 fat lambs or 1,000 store lambs at a time. He told us it was fairly quiet at that time as everybody was weaning lambs and shearing.

The next morning we made our way down to Invercargill to see the big sports complex where they were holding the shearing competition. We had booked a room in the hotel attached and had a look round to see where we were. Then we drove east along the south coast, which is great sheep country, and saw the best sheep. They were big ewes, Romney cross with clean heads, like a Lleyn, with good lambs. In some parts there were dairy cows, all colours of the rainbow. They cross the Friesian with Jersey and Ayrshire cows to improve the butter fat. We made our way to Lumsden where there was going to be a shearing competition the following Saturday. We called by and booked digs for two nights just outside the town. Then we then went up the east coast to visit the Dowling family again and stayed two nights there.

On the first morning Mark said he was short-staffed. His daughter was at college and his wife was away, but he had 800 lambs to spray that afternoon. We told him we would come back and give him a hand. We went off to look around the country and came back about 2.30pm. Mark had the lambs in, so when the sprayer came and set up, we helped,

Back to New Zealand

with Ray and me on the race, and Mark and four Huntaway dogs behind the sheep. They were noisy dogs, but they shifted the sheep well. The lambs had already been sheared and we put them through the sprayer in thirty minutes, so Mark was pleased. They were mostly Southdown cross lambs that would be forty to forty-five kilos when fat.

After that, Mark's brother rang on the mobile, wanting a hand to get 1,000 lambs across the main road. Mark said he wanted to get some more sheep in, so Ray and I went down and helped. We got them over and had a good chat with John. The lambs were going to chicory and red clover and would be fat in about six weeks. We stayed another night and moved on.

We were staying in South Island, so we travelled through Arthur's Pass, which is rough country, then back down the west side, visiting the glaciers again. They had retreated a lot in four years. We then moved down and called in a small town and stayed in a bed and breakfast. We went to eat in a pub just up the road where they had an open mike night and there was also free food, so we had a good night. The music was good, but the singing wasn't up to much!

Next morning we made our way down to Winton again and managed to get a room in the same place as before, she was very good to us. The next night there was a speed shear outside the back of a pub down the road, so we went along to see it. There was a lorry parked on one side of the park with curtain sides and a ramp to get the sheep up. The whole set-up worked well and the shearing was good.

At one point I was leaning on the wall, about ten yards back, and behind the wall was a gang of girls and some boys. They were telling jokes, so I asked them if they wanted to hear the latest Welsh joke. They said they did, so I told this joke: 'This lady went into a sex shop and asked the lady behind the counter could she direct her to the vibrator department. The shop lady put her hand up and waggled her finger and said

193

"Come this way." The customer said "If I could come that way, I wouldn't want a vibrator!"'

They laughed, and a minute later one of these girls leaned over and whispered in my ear, would I like a shag?! I couldn't believe a woman of about twenty was after a seventy-four-year-old man! She must have thought I knew what I was talking about! When I told Ray he said 'You lucky bugger!' I didn't take her up on the offer.

Next day we went across country to Lumsden to see the Welsh team competing, but they didn't get to the final. On the Sunday morning we made our way to Gorey for a speed shear and dog trial. We didn't see the dog trial that had been held in the morning. We found it funny, because they don't send the dog to fetch the sheep. They stand about thirty yards off the sheep pen, and when the sheep come out the dog starts working the sheep around different gates and races and the shepherd walks with the dog. It was interesting to see and there were some good dogs at work.

In the afternoon we went over to the hall in the middle of the park for the speed shear. It was supposed to start at 2pm and there were seventy-six shearers ready to go, but there was no sign of them starting even at 2.45pm. I said to Ray perhaps they didn't have a commentator, so I went across and asked one of the bosses did they need any help. He put the mike in my hand and said 'Get on with it!' I did the lot. I put seventy-six shearers through twice and the fastest ten again, so we had a final of three. The fastest man got down to 20.3 seconds. The Welsh boys came in the money and sheared well. Darren Ford, at the age of 51, sheared his lamb clean in 21 seconds. The winner's lamb wouldn't have got through in Wales, it had too much wool left on it.

There were about six people on the committee. No-one had asked a commentator to come, so they were lucky I was there.

We made our way back to Winton and stayed with the

Back to New Zealand

same lady once again. She was great; if she didn't have a spare room, she would make one up for us. There was a speed shear up the road the next night, and that night down the pub we met Julia Ford, Darren's wife. She asked us to her place for supper the following night after the sheep shear. 'It's only just down the road past the roundabout on the left,' she told us.

The next night we drove about twenty miles before we saw the roundabout and then missed the turn to her place, so we had to turn around and, on the way back, we spotted it hidden and set back from the road. It was a lovely place, a big four-bed bungalow with plenty of ground around it. We went to the shed where she was feeding the shearers and met Darren and the gang. Former team champion Hamish Mitchell from Scotland was there, too.

Julie told us to sit down and she would get us some food. She sliced us a plate of pork, half-an-inch thick, and we enjoyed the best meal we were served in New Zealand. After the meal Darren asked us into the bungalow and gave us a drink and we talked about shearing in between phone calls. He had a large sheet of paper on the table and was organising shearers to six different farms the next day. He had seven going to one, five to another and so on. He knew what number each shearer could do, and he would swap a fast shearer to make the numbers work on one farm, then take a slow one out to go somewhere else. It's a lot of work organising gangs like that, but he works with one gang himself, so he knows who does what and how tidy everyone is. He was a New Zealand record-holder on shearing ewes at one time, over 700 ewes. His brother, Edsel Ford, came to the Royal Welsh Show some years ago.

When we moved down to Invercargill for the World Championships we moved into our digs for eight days and met up with many people from all over Wales there for the championships. On the Tuesday, Ray had a World Shearing meeting, so I went down to the sports centre to watch the local shearing classes and saw some Welsh boys doing very well. I was sat by a retired chap who was an auctioneer and valuer in his younger days, and after half an hour he asked if Ray and I had digs, as we could stay with him, if not. I said we were sorted, but that's what New Zealanders are like.

That night there was a World Championships dinner at the hotel, so we went. It was a posh do and David Fagan was the guest speaker. He spoke very well and afterwards we made our way to the bar and enjoyed some good company.

The next day, the heats of the hand shearing and wool handling started – our teams did well. Huw Condron and Wyn Jones, from Wales, were commentating. Huw was doing some introducing and announcing, and he is very popular in New Zealand. When we got to the open heats, the New Zealanders were on their own sheep – and it showed. The Welsh team of Ian Jones and Gwion Evans are both good shearers, but Ian blew his wrist during the first days he was over there, so didn't get as much practise as he would have liked. That meant Gwion was carrying the team a bit, and they didn't make the final unfortunately.

The World Championships final was won by Johnny Kirkpatrick, a South Island shearer, with Gavin Mutch second and Jason Stratford third. Hamish Mitchell was the fastest man, passing Gavin Mutch on the sixteenth sheep. They had to shear five Merinos, five Romney, five crossbred and five lambs. Gavin had been in front until then, but his first lamb was a poor one and he slowed down a bit so as to do a good job, and Hamish passed him, going like hell. Gavin got back at him afterwards, although it upset his rhythm, but he still finished a sheep ahead of Johnny.

If Hamish hadn't been in the final, Gavin would have won. But we were in New Zealand and they wanted Johnny or Jason to win, as it was held on South Island. Hamish Mitchell finished fifth.

Back to New Zealand

On the last night there was a party at the hotel. We had bought tickets which included food, but the band was too loud and you couldn't hear yourself think. We took our food to the bar and enjoyed a pint or two with Alun McDonald, David Fagan, his brother John, and the rest of the team that organised the championships.

The crowd in the sports hall was about 4,000, which they said was the biggest ever to watch the World Championships. We had 6,000 to 7,000 at the Royal Welsh Show in 2010.

We moved out the next morning and headed for Queenstown to meet our plane to take us to Australia and then home. It takes four and a half hours to fly over to Australia. They had suffered bad bush fires in South Australia, and we could still see a lot of burning from the plane. We arrived at the airport in a terrible thunderstorm and had to wait to disembark.

We also had hours to wait for the next plane, so we had some food and a walk around, as the next part of the journey was fourteen hours to Dubai, a long haul. The next plane was a big jet and we tried to upgrade, but they wanted £1,100 extra, so we stuck to our own seats. I was next to the window, so enjoyed going over places like Iraq, but was hoping they wouldn't start shooting at us! Ray wasn't sat by me, so after about ten hours he came round to see if I was OK. I said I was, and that we wouldn't be too long now. 'What are you on about?' he said. 'It's another four hours!' I couldn't believe it, my backside was numb by then!

Abu Dhabi didn't have a very good airport – there were too many people in a small place, but it was better on the way back. We had a short break and got on the next plane – another seven hours. The food on the plane wasn't very good. We travelled with Air Arabia, and they served rice with everything, but we managed.

CHAPTER 33

Home... and alone

I WAS GLAD to be home at last. I don't suppose I will ever go again, but I loved New Zealand and the people looked after us well. We made a lot of new friends who come over to the Royal Welsh Show.

Before going, my wife had a cold and cough and when I arrived home she was still coughing. She had taken antibiotics, but it didn't stop. The first job was to go for an X-ray. This showed up some problems, so we went to Hereford for a scan and the following week we met the specialist who told us it looked like lung cancer. This was a shock, as Marg had never smoked in her life and nor had I. He said forty per cent of non-smokers got cancer, with sixty-seven per cent of smokers getting it. After that, she had all sorts of tests and samples taken from the lung, which was not very nice. Thank God I had retired, so I had plenty of time to take Marg to Hereford hospital. We spent a lot of time there and met a lot of other people with cancer.

Marg started taking a tablet, but this only worked for a few months. This was then changed to a new immunotherapy drug. It worked alright, except that Marg was coughing worse than ever, so they decided she could have some radiotherapy. This worked well and stopped the cough, which was a big help. So every three weeks she had another dose of immunotherapy, and a scan every seven or eight weeks to check whether it was still working – and we hoped for the best.

She said she was ready for whatever happened, but

Home... and alone

it scared me. We had been together for nearly sixty years between courting and getting married.

At the 2015 Royal Welsh Show I had been presented with Royal Welsh Honorary Life Governorship for services to shearing. At the 2017 show I was presented with the Certificate of Fellowship, making me a Fellow of the Royal Agricultural Society. I had been an Associate Member for five years, so this was my Welsh MBE.

That brought me to the age of seventy-five. I had the honour of judging the Beulah Speckled class at the Royal Welsh Show in 2017, and judged in Hundred House Show in September and Llantrisant Show in August. The end of October saw me judging the overall champion of the day in Cambridge at the British Hedging Championships – that was out of 120 hedgers of all styles. It was the first time I'd judged the overall competition, though I had judged the Welsh styles a few times over the years.

It was a year of ups and downs in 2019. Marg was struggling with the cancer and we were travelling to Hereford hospital for chemotherapy every two or three weeks, with scans in between. She would suffer badly for four or five days after the chemo, but once she got over that she was pretty good.

During the first week of June I had a shock when I received a letter from Buckingham Palace telling me that I was to be presented with the MBE for 'Services to Farming Heritage'. Marg told me *she* should have it for keeping the diary and getting me to the many meetings I did over the years! We didn't hear any more about it until the first week in August, when we had the invitation and the paperwork telling me I was to be presented on 5 February 2020, at 10.30am in Buckingham Palace. Marg was very disappointed it was so far in the future, as she was getting worse. I had to fill in a form saying who would accompany me on the day. I was allowed three, so I put Marg's name down, but I was not sure she would be there the way things looked. I also put Amanda and Michael's names down to make my three people.

Over the summer I managed to judge the overall sheep classes at two shows, but failed to do my commentary at Brecon and Kington shows as Marg was too ill to leave her. When we got into September, Marg went downhill and spent two spells in Hereford hospital. The last time she was there she had a seizure which didn't do her any good. She came out of it but could no longer have any more treatment. We said we wanted her to come home and they said that was for the best. We got a proper hospital bed and oxygen in place, so everything was ready and home she came. She had good days and bad days but between all of us we managed alright.

The final week was tough. Marg was getting worse daily. On the afternoon of 25 September, I was sitting with her, and a couple of times I thought she was going to go but she had rallied again by teatime.

The next day she looked awful but was able to talk a little bit. I asked was there anything she wanted. She said she wanted to see her brother Colin, and also Sally and Jeff from Tonteg. Sally and Marg were like sisters. Amanda came to sit with Marg and I phoned Colin and told him things were bad. He said he would be able to be with us by 5pm. I contacted Sally and Jeff, who were also going to come by 5pm.

When they arrived Colin went in to see Marg and I went to make some tea. When he came out he said she had talked a bit but had gone to sleep for a while. After their tea, Sally and Jeff went in to see Marg as she was awake again by then, and she talked a little but was very weak. Colin left about 6.30pm to drive back to Bradford, while Sally and Jeff went back about 7pm.

By now, Michael and his wife Julie had arrived. I told Julie, who is a nurse, that things did not look good and to prepare for the worst. Marg went downhill from then on and finally, at twenty past ten, she passed away. We were all there. It was

Home... and alone

very sad. I don't wish that on anyone. We had been married fifty-five-and-a-half years and courting for four before that.

Then we had to think about the funeral. We were lucky the undertaker was free on 5 October. I went to see Dave, the landlord of the Trout Inn, who was also free, so that was good. Then there were flowers, leaflets and hymns to think about. Marg had written her eulogy and chosen Estyn James to read it, so she saved me a lot of work. She wanted Tom Jones singing 'Green, Green, Grass of Home' on the way out of the church. I had chosen 'Wind Beneath my Wings' by Daniel O'Donnell on the way in and during the service David Price sang 'The Rose'.

The next day the cards started arriving. By the end, I had run out of room to put them all. I had over 200, and Amanda and Mike also had a lot too.

On the morning of the funeral the flowers arrived and they were beautiful. When we arrived at the church there crowds of people from all over, but when we came out of the church, walking across to the grave was tough for me. That was the worst part of the day, seeing the coffin going into the ground, but you have to face it.

I shook hands with all of the bearers and thanked them. I then moved on to the crowd. The first person I met was Ari Ashley from Kent, who used to live at Ty'n y Cwm, and was in floods of tears. I then went with the rest of the family down to the gate. I've never shaken so many hands ever – it was one of the biggest funerals held at Beulah church. Marg would not have believed how many people turned up. We were very grateful for all of the donations for Macmillan, the church and the Air Ambulance, which totalled over £3,500.

After meeting everybody we made our way down to the Trout Inn for tea – by then, I was ready for some. The pub was full. Mike had opened up the field opposite for parking and that was full too, as well as the pub car park and both sides of the road. After a while people drifted away, and we had room to sit down and have a few drinks and some food as a family.

After it was all done, I was left alone in the bungalow. That's when it's hard, but, having lived on my own for twelve months before we got married, I soon settled into a life of my own.

Christmas came and went and the family looked after me well. The weather was very wet, but later I did some hedging to keep my hand in.

CHAPTER 34

Life goes on

ON 4 FEBRUARY 2020 we travelled up to London to receive my MBE. We stayed the night and were at Buckingham Palace at 9.45am. There was a lot of fuss with the security, but we got in. When it was my turn to meet Prince Charles, as he was then, I had some fun because he is a hedger. So when my name was called I told him that I hadn't expected to see him that morning as it was lovely and dry. 'I thought you would be hedging,' I said. He laughed at that and asked whether I was still hedging. I said I was and he wanted to know what style. He said he liked the Welsh style of hedging. He pinned the MBE on my lapel and that was that. I met a lot of very nice people that day and had a very enjoyable time.

Just a month after that we ran into the start of the Coronavirus pandemic, which was worse than anything we had seen before. The whole world was affected and the loss of life was dreadful. Marg would have been petrified and her friend Sally lost her husband Jeff to the virus. A summer of no Royal Welsh or any other shows followed. Everything was cancelled, it was a long old summer.

After Sally lost Jeff, I phoned her most nights to try and keep her going, as they had phoned me after Marg had died. In the December, Sally was not well, and one night she said to me she had been having tests for cancer. She had the results the next day. I asked her how bad it was. She said she had cancer on both lungs and in one breast; it had also gone up her neck. I was shocked and told her she had a fight on her hands. She soon started chemotherapy. She had several lots and was doing well. I talked to her every night to try and keep her spirits up.

Sally finished one particularly long stretch of chemo and had a rest for a week. She was good for a few days but, by the Saturday, she was feeling rough again. I could tell on the phone that she was having a bad day. On the Sunday she was worse, and on the Monday she could hardly talk on the phone. I told her she should be in hospital, and she said they were taking her in the next morning. They did, but she died from sepsis that evening. I was shocked when they phoned to tell me. Amanda, Phil and I went to the service and to the golf club later for tea with her family and friends.

With the lockdowns, it wasn't possible to do much, but time went by and it was lambing again. I helped Mike with that – the days went by much faster when I was busy.

There was no Royal Welsh or Three Counties shows for a second year, so no commentating for me, and there were also no small shows again.

When autumn came the sheep sales went ahead, so that took some time out, but all the ploughing matches were off and there was no hedging to judge, either.

In the spring the lockdown eased a bit and it was possible to hold a meeting if people sat two metres apart and wore a mask.

The Royal Welsh Show held a meeting to present many people with medals for long service. I received a Silver Medal and Lifetime Membership to mark over forty years' work for the show. I had been commentating and shearing for forty-three years, and a member of the shearing committee for that time as well.

At the Three Counties Show they presented me with an engraved whisky decanter for over forty years of commentating there. It's a great show, and over the years the shearing has improved a lot. Years ago, Godfrey Bowen

Life goes on

would come and watch the shearing on the Saturday and the Open final in the afternoon. I would go down and ask him if he would like to commentate on the final with me. He would always say yes and enjoyed it very much.

And now? I still do a bit of hedging to keep my hand in and earn some money, but I'm finding it harder as my age catches up with me. Hedging has always been part of my life and I am still judging. Last autumn I judged the overall champion up in Hafren, near Llanidloes. I have made many friends in the hedging world and enjoy going to the matches, but it's going to get harder to keep the matches going as very few young people are taking it up.

To get some fresh air, I get my 4x4 bike out and go for a ride up Ty'n y Cwm to see how my grandson Paul is farming. I usually find something wrong for him or Mike as I go around – sheep in the wrong place or gates being left open – we have a lot of walkers going through.

It's a great view from the top of Ty'n y Cwm, out over Llofftybardd Farm and Pantycelyn chapel and up the valley to Trallwm. I always enjoy my ride up around there and hope to do so for many a year to come.

As I close this book, I would like to thank my family and friends for all their support over the years.

Also from Y Lolfa:

Life Beneath The Arch
CHARLES ARCH

£9.99

A Life to Dai For

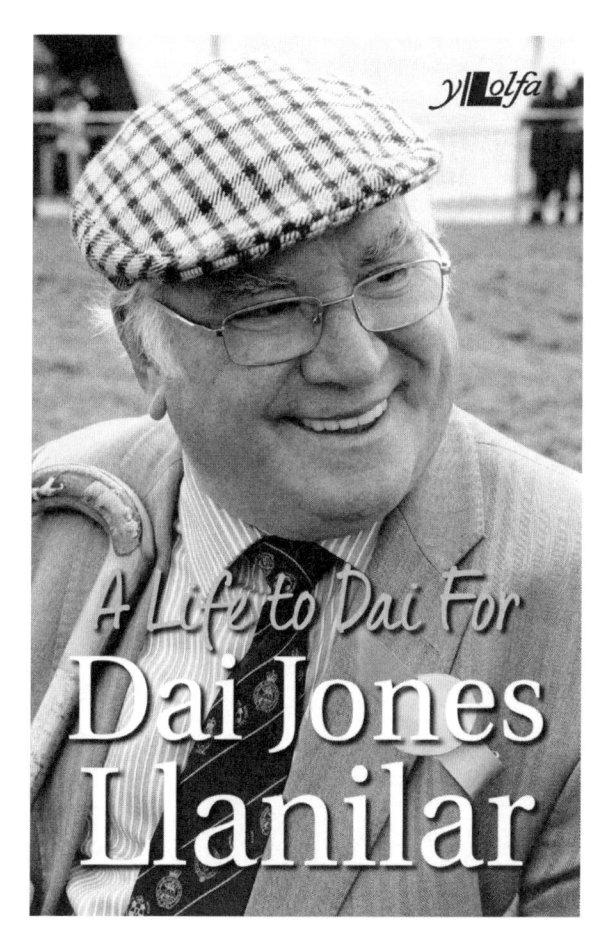

Dai Jones
Llanilar

yLolfa

£9.99